THE
SPIKE

How Our Lives Are Being Transformed
by Rapidly Advancing Technologies

Damien Broderick

A TOM DOHERTY ASSOCIATES BOOK
NEW YORK

THE SPIKE: HOW OUR LIVES ARE BEING TRANSFORMED BY RAPIDLY
ADVANCING TECHNOLOGIES

This book is printed on acid-free paper.

Edited by David G. Hartwell

Book design by Jane Adele Regina

Graph by Anders Sandberg (asa@nada.kth.se)

A Forge Book
Published by Tom Doherty Associates, LLC
175 Fifth Avenue
New York, NY 10010

www.tor.com

Forge® is a registered trademark of Tom Doherty Associates, LLC.

Library of Congress Cataloging-in-Publication Data
Broderick, Damien.
    The spike : how our lives are being transformed by rapidly advancing
technologies / Damien Broderick
        p. cm.
    "A Tom Doherty Associates book."
    Includes bibliographical references.
    ISBN 0-312-87781-1 (hc)
    ISBN 0-312-87782-X (pbk)
    1. Technology—Social aspects. 2. Technological forecasting I. Title
T14 .B75 2001
303.48'3—dc21   00-049043

First Hardcover Edition: February 2001
First Trade Paperback Edition: February 2002

Printed in the United States of America

0 9 8 7 6 5 4 3 2 1

"Broderick treats technological change as the next step in human evolution and provides a compelling version of how we stand to change as a species in the next few years. Fascinating stuff from a capable writer of popular science."

—*Booklist*

"*The Spike* is the most exciting, provocative book of the year. . . . His dispatches from the way-out frontier of speculative research are exhilarating for the sheer marvel of what might be."

—*Microsoft Communiqué*

"Broderick [is] a seasoned and respected interpreter of science, whose strength is in seeing the connection between otherwise widely separated areas of knowledge."

—*The Age* (Melbourne, Australia)

"Since this chapter was written, these ideas have been developed in great detail by such writers . . . [as] Damien Broderick (*The Spike*, 1997). Damien's book will serve as a more imaginative sequel to the one you are reading now."

—Arthur C. Clarke, in the revision of his classic *Profiles of the Future*

"A wide-ranging tour of some speculative and often exciting areas of technological development . . . I recommend *The Spike* to anybody deeply interested in the future. . . . As a guide to the technologies of tomorrow, the book makes a stimulating and provocative read. Warmly recommended."

—Dr. Michael Nielsen, *Extropy Online*

"A provocative account of cutting-edge science and technology that deserves a place on the desks of political leaders and public policy makers everywhere. . . . Things are happening incredibly faster than most of us think, and the winners are going to be those who get used to the idea and work out in advance what has to be done in order for us to cope."

—Dr. Race Matthews, *The Age* (Melbourne, Australia)

"In *The Spike,* the science and technology rushing history towards its 'singularity' and beyond are presented by Broderick in a masterpiece of clarity."

—Alan Olding, *The Australian's Review of Books*

To Vernor Vinge and Hans Moravec,

who galloped ahead up the Spike's slope
to fetch back the strange news

Contents

THE
SPIKE

1: The Headlong Rush of Time

IF OUR WORLD SURVIVES, THE NEXT GREAT CHALLENGE TO WATCH
OUT FOR WILL COME—YOU HEARD IT HERE FIRST—WHEN THE CURVES
OF RESEARCH AND DEVELOPMENT IN ARTIFICIAL INTELLIGENCE, MO-
LECULAR BIOLOGY, AND ROBOTICS ALL CONVERGE. ODOY. IT WILL BE
AMAZING AND UNPREDICTABLE, AND EVEN THE BIGGEST OF BRASS, LET
US DEVOUTLY HOPE, ARE GOING TO BE CAUGHT FLAT-FOOTED. IT IS
CERTAINLY SOMETHING FOR ALL GOOD LUDDITES TO LOOK FORWARD
TO IF, GOD WILLING, WE SHOULD LIVE SO LONG.
　　—THOMAS PYNCHON, *NEW YORK TIMES BOOK REVIEW*, 1984[1]

WITHIN THIRTY YEARS, WE WILL HAVE THE TECHNOLOGICAL MEANS
TO CREATE SUPERHUMAN INTELLIGENCE. SHORTLY AFTER, THE HU-
MAN ERA WILL BE ENDED.
　　—VERNOR VINGE, NASA VISION-21 SYMPOSIUM, 1993[2]

It rushes at you, the future.

Usually we don't notice that. We are unaware of its gallop.
Time might not be a rushing black wall coming at us from the
future, but that's surely how it looks when you stare unflinchingly
at the year 2050 and beyond, at the strange creatures on the near
horizon of time (our own grandchildren, or even ourselves, technologically preserved and enhanced). Call them *transhumans* or
even *posthumans*.

The initial transition into posthumanity, for people intimately
linked to specially designed computerized neural nets, might not
wait until 2050. It could happen even earlier. Twenty-forty.
Twenty-thirty. Maybe sooner, as Vinge predicted. This is no
longer the deep, the inconceivably distant future. These are the
dates when quite a few young adults today expect to be packing

up their private possessions and leaving the office for the last time, headed for retirement. These are dates when today's babes in arms will be strong adults in the prime of life.

Around 2050, or maybe even 2030, is when a technological Singularity, as it's been termed, is expected to erupt. That, at any rate, is considered opinion of a number of informed if unusually adventurous scientists. Professor Vinge called this projected event "the technological Singularity," something of a mouthful. I call it "the Spike," an upward jab on the chart of change, a time of upheaval unprecedented in human history.

And, of course, it's a profoundly suspect suggestion. We've heard this sort of thing prophesied quite recently, in literally Apocalyptic religious revelations of millennial End Time and Rapture.

That's *not* the kind of upheaval I'm describing.

A number of perfectly rational, well-informed, and extremely smart scientists are anticipating a Singularity, a barrier to confident anticipation of future technologies. I prefer the term *Spike*, because when you chart it on a graph it *looks* like a Spike! Its exponential curve resembles a spike on a graph of change over time. Here's a picture of it:

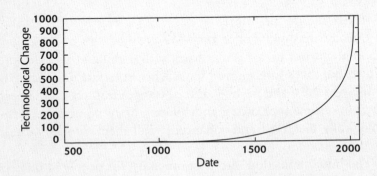

As you see, the more the curve grows, the larger is each subsequent bound upward. It takes a long time to double the original value, but the same period again gets you four times farther up the curve, then eight times . . . so that after just *ten* doublings, you've risen a *thousand* times as far, then two thousand, and on it goes. Note this: the time it takes to go from one to two, and then from two to four, is just the same period needed to take that mighty leap from 1000 to 2000. A short time later we're talking a millionfold increase in a single step, and the very next step after that is *two millionfold* . . . [3]

History's slowly rising trajectory of progress over tens of thousands of years, having taken a swift turn upward in recent centuries and decades, quickly roars straight up some time after 2030 and before 2100. That's the Spike. Change in technology and medicine moves off the scale of standard measurements: it goes *asymptotic*, as a mathematician would say. An asymptote is a curve that bends more and more sharply until it is heading almost straight along one of the axes—in this case, up the page into the future.

So the curve of technological change is getting closer and closer to the utterly vertical in a shorter and shorter time. At the limit, which is reached quite quickly (disproving Zeno's ancient paradox about the tortoise beating Achilles if it has a head start), the curve tends toward infinity. It rips through the top of the graph and is never seen again.

At the Spike, we can confidently expect that some form of intelligence (human, silicon, or a blend of the two) will emerge at a posthuman level. At that point, all the standard rules and cultural projections go into the waste-paper basket.

A quick preliminary stroll through the future

Everything you think you know about the future is wrong.

How can that be? Back in the 1970s, Alvin Toffler warned of future shock, the concussion we feel when change slaps us in the back of the head. But aren't we smarter now, in the twenty-first century? We have wild, ambitious expectations of the future, we're not frightened of it. How could it surprise us, now that *Star Trek* and *Star Wars* and *Terminator* movies and *The Matrix* and a hundred computer role-playing games have domesticated the twenty-fourth century, cyberspace virtual realities, and a galaxy far, far away?

Actually, I blame glitzy mass-market science fiction script writers for misleading us. They got it so wrong. Their enjoyable futures, by and large, are about as plausible as nineteenth-century visions of tomorrow. Those had dirigibles filling the skies and bonneted ladies in crinolines tapping at telegraphs.

Back in the middle of the twentieth century, when the futuristic stories I read as a kid were being written, most people knew "that Buck Rogers stuff" was laughable fantasy, suitable only for children. After all, it talked about atomic power and landing on the moon and time travel and robots that would do your bidding even if you were rude to them. Who could take such nonsense seriously?

Twenty years later, men *had* walked on the moon, nuclear power was already obsolete in some countries, and computers could be found in any university. Another two decades on, in the nineties, probes sent us vivid images from the solar system's far reaches (and got lost on Mars), immensely powerful but affordable personal computers sat on desks at home as well as work, the human genome was being sequenced, and advanced physics told us that even time travel through spacetime wormholes was not necessarily insane (although it was surely not in the immediate offing).

So popular entertainment belatedly got the message, spurred on by prodigious advances in computerized graphics. Sadly, the movie, television, and game makers still didn't know a quark from a kumquat, a light-year (a unit of interstellar distance) from a picosecond (a very brief time interval). With gusto and cascades of light, they blended made-up technobabble with exhilarating fairy stories, shifting adventure sagas from ancient legends and myth into outer space. It was great fun, but it twisted our sense of the future away from an almost inconceivably strange reality (which is the way it will actually happen) and back into safe childhood, that endless temptation of fantastic art.

Maybe you think I'm about to get all preachy and sanctimonious. You're waiting for the doom and gloom: rising seas and greenhouse nightmare, cloned tyrants, population bomb, monster global megacorporations with their evil genetically engineered foods and monopoly stranglehold on the crop seeds needed by a starving Third World. Wrong. Some of those factors indeed threaten the security of our planet, but not for much longer (unless things go very bad indeed, very quickly). No, what's wrong with most media images of the future isn't their evasion of such threats—on the contrary, they play them up to the point of absurdity. What's wrong is their laughably timid *conservatism*.

The future is going to be a fast, wild ride into strangeness. And many of us will still be there as it happens.

That strangeness is exactly what prevents us from picking out any one clear determinate future. The coming world of the Spike is, strictly, unimaginable—but we can certainly try our best to trace some of the contributing factors, and some of the ways they'll converge (or perhaps block each other). That fact governs my approach in this book. Do not expect a dogmatic manifesto advancing a single thesis. Instead, I'll try to give you a glimpse of many different technologies. I won't attempt the impossible, which is to integrate all those different points of view into one comforting, assured framework. There is no inevitable tomorrow.

All that we know for sure is the almost unstoppable acceleration of science and technology, and the drastic impact it will have upon humanity and our world.

Living in the future right now

This accelerating world of drastic change won't wait until, say, *Star Trek*'s twenty-fourth century, let alone the year 3000. We can expect extraordinary disruptions within the next half century. Many of those changes will probably start to impact well before that. By the end of the twenty-first century, there might well be no humans (as we recognize ourselves) left on the planet—but, paradoxically, nobody alive then will complain about that, any more than we now bewail the loss of Neanderthals.

That sounds rather tasteless, but I mean it literally: many of

us will still be here, but we won't be human any longer—not the
current model, anyway. Our children, and perhaps we as well,
will be smarter. We already have experimental hints of how that
might occur. In September 1999, molecular biologists at Princeton reported adding a gene to a strain of mice, elevating their
production of NR2B protein. The improved brains of these "Doo-
gie mice" used this extra NR2B to enhance brain receptors, helping the animals solve puzzles much faster. A kind of genetic
turboaccelerator for mousy intelligence. Human brains, as it happens, use an almost identical protein. It is not far-fetched to suppose that we will learn to tweak or supplement it to increase our
own effective intelligence (or that of our children).

Nor will we be the only high-level intelligences on the planet.
By the close of the twenty-first century, there will be vast numbers
of conscious but artificial minds on earth. How we and our children get along with them as they move out of the labs and into
the marketplace will determine the history of life in the solar
system, and maybe the universe.

I'm not making this up. Dr. Hans Moravec, a robotics pioneer
at Carnegie Mellon University in Pittsburgh, argues in *Robot*
(1999) that we can expect machines equal to human brains within
forty years at the latest. Already, primitive robots operate at the
level of spiders or lizards. Soon a robot kitten will be running
about in Japan, driven by an artificial brain designed and built
by Australian Dr. Hugo de Garis. True, it's a vast leap from lizard
to monkey and then human, but computers are *doubling* in speed
and memory *every year*.

This is the hard bit to grasp: with that kind of annual doubling
in power, you jump by a factor of 1000 every decade. In twenty
years, the same price (adjusted for inflation) will buy you a computer a *million* times more powerful than your current model.
That's "Moore's law," enunciated in 1965 by Gordon E. Moore,
one of the founders of the Intel company, which now makes the
Pentium chip for your personal computer. Moore originally sur-
mised that the number of components on an integrated circuit
(IC) would double each year. If that were to happen, 65,000 transistors would dance on an IC within ten years. That was a little

ambitious, but turned out to be close to reality—a result nobody could have believed in 1965. Moore's conjecture changed as time passed, first slowing down to "doubling every two years" then speeding back up to "doubling every eighteen months." It remained an astonishing prediction, and an amazing phenomenon.

Moore's law (although of course it isn't really anything like a law of nature) made a disarmingly simple algebra equation, the kind even someone uncomfortable with figures might be able to follow: doubling goes as *two raised to the power of N*, and currently "N" is conjectured to equal "one and a half years." With this equation you can work out how fast it takes to get to a millionfold increase, say, by following Moore's (revised) law through the following simple steps:

- two to the tenth power equals roughly 1000
- two to the twentieth power equals a million
- and two to the fortieth power equals a thousand billion.

If, to be conservative, a single doubling happens during each two-year period, then every twenty years we get a thousand times as much computational power per dollar as we started with.

At the start of the 2000s, the world's best, immensely expensive supercomputers perform several trillion operations a second. To emulate a human mind, Moravec estimates, we'll need systems a hundred times better. Advanced research machines might meet that benchmark within a decade, or sooner—but it will take another ten or twenty years for the comparable home machine at a notepad's price. Still, around 2040, expect to own a computer with the brain power of a human being. And what will *that* be like? If software develops at the same pace, we will abruptly find ourselves in a world of alien minds as good as our own.

Will they take our orders and quietly do our bidding? If they're designed right, maybe. But that's not the kicker. That's just the familiar world of third-rate sci-fi movies with clunky or sexy-voiced robots. The key to future change comes from what's called "self-bootstrapping"—machines and programs that modify their own design, optimize their functioning, improve themselves in ways that limited human minds can't even start to understand. Dr. de Garis calls such beings "artilects," and even though he's building their predecessors he admits he's scared stiff.

By the end of the twenty-first century, computer maven Ray Kurzweil (in *The Age of Spiritual Machines*, 1999) expects a merging of machines and humans, allowing us to shift consciousness from place to place. He's got an equally impressive track record, as a leading software designer and specialist in voice-activated systems. His time line for the future is even more hair-raising than Moravec's. In a decade, he tells us, expect desktop machines with the grunt of today's best supercomputers, a trillion operations a second. Forget keyboards—we'll speak to these machines, and they'll speak back in the guise of plausible personalities.

By 2020, a Pentium equivalent will equal a human brain. And now the second great innovation kicks in: molecular nanotechnology (MNT), building things by putting them together atom by atom. I call that "minting," and the wonderful thing is that a mint will be able to replicate itself, using common, cheap chemical feedstocks. Houses and cars will be compiled seamlessly out of diamond (carbon, currently clogging the atmosphere) and sapphire (aluminum), because they will be cheap appropriate materials readily handled by mints.[4] It's not clear, however, if one-size-fits-all universal assemblers will be feasible, at least in the near future; some mints might be specialized to compile carbon compounds, others to piece together aluminum (into sapphire) or tungsten-carbide structures, requiring assembly at a coarser level. These dedicated mints will operate at successively higher temperatures, each requiring a totally different chemistry (feedstock, tool-tips, energy sources). "Whether you can use one level of MNTing to enable the next higher level," notes one commentator, "remains a very open question."[5]

Until recently, all nanotechnology was purely theoretical. A Rand Corporation study declared cautiously: "Extensive molecular manufacturing applications, if they become cost-effective, will probably not occur until well into the far term. However, some products benefiting from research into molecular manufacturing may be developed in the near term. As initial nanomachining, novel chemistry, and protein engineering (or other biotechnologies) are refined, initial products will likely focus on those that substitute for existing high-cost, lower-efficiency products."[6] The engineering theory was good, but the evidence was thin. Finally, though, at the end of November 1999, came a definitive breakthrough, harbinger of things to come. Researchers at Cornell University announced in the journal *Science* that they had successfully assembled molecules one at a time by chemically bonding carbon monoxide molecules to iron atoms. This is a long way from building a beefsteak sandwich in a mint the size of a microwave oven powered by solar cells on your roof (also made for practically nothing by a mint), but it's proof that the concept works.

If that sounds like a magical world, consider Kurzweil's 2030. Now your desktop machine (except that you'll probably be wearing it, or it will be built into you, or you will be absorbed into it) holds the intelligence of one thousand human brains. Machines are plainly people. It might be (horrors!) that smart machines are debating whether, by comparison with their lucid and swift understanding, *humans* are people! We had better treat our mind children nicely. Minds that good will find little difficulty solving problems that we are already on the verge of unlocking. Cancers will be cured, along with most other ills of the flesh.

Aging, and even routine death itself, might become a thing of the past. In October 1999, Canada's Chromos Molecular Systems announced that an artificial chromosome inserted into mice embryos had been passed down, with its useful extra genes, to the next generation. And in November 1999, the journal *Nature* reported that Pier Giuseppe Pelicci, at Milan's European Institute of Oncology, had deactivated the *p66shc* gene in mice—which then lived thirty percent longer than their unaltered kin, without making them sluggish! A drug blocking *p66shc* in humans might have a similar life-extending effect.

As well, our bodies will be suffused with swarms of medical and other nano maintenance devices. The first of three magisterial volumes detailing how and why medical nanorobots are in mid-range prospect appeared at the end of 1999: Dr. Robert A.

Freitas Jr.'s *Nanomedicine*. Nor will our brains remain unaltered. Many of us will surely adopt the prosthetic advantage of direct links to the global net, and augmentation of our fallible memories and intellectual powers. This won't be a world of Mr. Spock emotionless logic, however. It is far more likely that AIs (artificial intelligences) will develop supple, nuanced emotions of their own, for the same reason we do: to relate to people, and for the sheer joy of it.

The real future, in other words, has already started. Don't expect the simple, gaudy world of *Babylon-5* or even *eXistenZ*. The third millennium will be very much stranger than fiction.

Walking into the future

To get a firmer idea of the reasoning that underlies these apparently reckless claims, consider the ever-accelerating rate at which people have been able to travel during the last three hundred thousand years (or the last three million, if you're willing to accept a generous definition of humankind).

For very much the largest part of that span, we were limited to walking pace, with long rests. Some six thousand years ago we borrowed the lugging power of asses, then the strength and endurance of other large animals, finally coupling small ponies to war chariots in the second millennium B.C. Breeding horses large enough to ride took many centuries more. In other forms of transport, dugout canoes, then boats, and finally ships with sails went as fast as arms could paddle, or winds, captured fairly inefficiently, blow.

Less than two hundred years ago, steam trains sent our ancestors hurtling on rail at twenty or thirty kilometers per hour. Cars made faster speeds commonplace within the living memory of the elderly, especially as roads improved (at prodigious cost, financially and to the shape of the landscape). Prop aircraft flew at a few hundreds of kilometers per hour. Within decades, jets flew ten times that fast, and by the 1960s rockets took astronauts into space at tens of thousands of kilometers per hour. Today, using "virtual presence" on-line simulation systems, we are on the verge of "being there" (in a limited but vivid and interactive sense) at the speed of light. And that's the end of the line—you can't get faster than the velocity of light.

Mapped on a graph, this progression shows a long flat rise, turning slowly upward, then climbing more sharply, and faster again . . . and now its dotted projection seems to soar dizzyingly toward a veritable Spike.

The brain on your desk

That same headlong acceleration applies, as we are now uncomfortably aware, with the speed, power, and cheapness of computers. Computer-power-per-dollar currently doubles every eighteen months, or perhaps as swiftly as every year. Growth in computing power is already exponential, maybe *hyper*exponential.

Starting small, with one or two special highly secret vacuum-tube computers during the Second World War, the computer presence sluggishly increases to bulky, cantankerous devices in a few rich universities, and then some clumsy IBM mainframes in large businesses, and then the big vulnerable tubes get replaced by transistors, by integrated circuits, and before you know where you are it's the late 1970s, early 1980s, and home enthusiasts are buying their first Macs and PCs, and the prices continue to fall, and meanwhile the military and NASA are funding superfast giant machines running in a bath of liquid helium to keep them cool enough to function, and the curve is getting steeper and steeper—

It is the fable of the Chessboard brought to life: one grain on the first square, two on the second, four on the third—and by the time we reach the sixty-fourth square, we groan beneath a deluge of rice.

Computing power that is developing with such acceleration may be able to emulate human intelligence within thirty or forty years. A century, tops.

At that point, if the chart of the Spike is telling us the truth, we (or our children, or our grandchildren) may see machines with *twice* our capacity within a further eighteen months, then *four* times our capacity within a further year and a half, and . . .

Intelligence will have Spiked. It won't be *our* human intelligence, but we will be borne along into the Spike with it.

Spikes everywhere

Computing power and speed of travel are just two examples of runaway progress. The most exciting prospect, one that convinces scientists who assess the evidence for a coming Spike, is that other disciplines will have Spiked at about the same time: medical research into aging, cloning and genome manipulation, miniaturization of high-tech products until they reach molecular or even atomic scales (nanotechnology), and more.

So the world of the Spike will be marked by

- augmented human abilities, made possible by connecting ourselves to chips and neural networks that are not in themselves aware but can amplify our native abilities . . .
- human-level Artificial Intelligences (AIs), swiftly followed by *hyper*intelligent AIs . . .
- DNA genome control, which gives us the capacity to re-design ourselves and our children, enhancing not just mind but every bodily and emotional pleasure and aptitude . . .
- nanotechnology machines, including AIs, built from the atom up, including extremely tiny self-replicating devices no larger than molecules . . .
- extreme physical longevity or even (in effect, barring accident) immortality, due to a blend of: the new understanding and control of our genetic inheritance, including apoptotic "suicide genes" that may limit lifespan by restricting the number of times most cells can be repaired by self-replication; nanotechnological medical repair sys-

tems that live inside the body from birth and keep cells rejuvenated and free of disease, including cancers; "backup" copies of our memories maintained in machine storage in case of damage to the brain, or permitting organ or tissue cloning and replacement of lost knowledge and experience in the extreme case of severe physical damage to the body/brain . . .

- "uploads" or transfers of human minds into computers, so that we can live, work, and play inside their rich and manipulable machine-generated virtual realities . . .
- possible contact with galactic civilizations that have already gone through the Spike transition, including such extreme prospects as ancient extraterrestrial cultures so powerful that they have long ago restructured the visible universe (or rewritten the laws of quantum mechanics) . . .

First glimpses of the singularity

The core notion in these forecasts was first described metaphorically as a technological Singularity (although others had anticipated the insight, as we shall see in a moment) by Professor Vernor Vinge, a mathematician in the Department of Mathematical Sciences, San Diego State University. Why this curious and unfamiliar term "singularity"? It's a mathematical point where analysis breaks down, where infinities enter an equation. And at that point, mathematics packs it in.[7] A black hole in space is a kind of spacetime example of this rather abstract pathology. Hence, cosmic black holes, those ultimate mysteries with interiors forever beyond our exploratory reach, are also known as "singularities." "The term 'singularity' tied to the notion of radical change is very evocative," Vinge told me, adding: "I used the term 'singularity' in the sense of a place where a model of physical reality fails. (I was also attracted to the term by one of the characteristics of many singularities in General Relativity—namely the unknowability of things close to or in the singularity.)"[8]

For Vinge, accelerating trends in computer sciences converge somewhere between 2030 and 2100 to form a wall of technological novelties blocking the future from us. However hard we try, we cannot plausibly imagine what lies beyond that wall. "My 'technological singularity' is really quite limited," Vinge told me. "I say that it seems plausible that in the near historical future, we will cause superhuman intelligences to exist. Prediction beyond that point is qualitatively different from futurisms of the past. I don't necessarily see any vertical asymptotes."[9] So enthusiasts for this perspective (including me) are taking the idea much further than Vinge. Humanity, it is argued, will become first "transhuman" and then "posthuman." Under either interpretation, and unlike many currently fashionable debates, Vinge's singularity is an apocalyptic prospect based on testable *science* rather than religion or New Age millenarianism.

While Vinge first advanced his insight in works of imaginative fiction, he has featured it more rigorously in such formal papers as his address to the Vision-21 Symposium, sponsored by NASA Lewis Research Center and the Ohio Aerospace Institute, 30–31 March 1993. Professor Vinge opened that paper with the following characteristic statement, which can serve as a fair summary of my own starting point:

"The acceleration of technological progress has been the central feature of this century. I argue in this paper that we are on the edge of change comparable to the rise of human life on Earth. The precise cause of this change is the imminent creation by technology of entities with greater than human intelligence."

This remarkable prospect has not gone unnoticed among academics. University of Washington neurophysiologist Dr. William Calvin has discussed Vinge's claims in a vivid paper, "Cautions on the Superhuman Transition." Experts who have addressed the Singularity as specialists in fields such as AI research and nanotechnology include Dr. Hans Moravec, director of the Carnegie Mellon Mobile Robot Lab, Dr. Eric Drexler, director of the Foresight Institute, and Dr. Ralph Merkle, principal research fellow at Jim von Ehr's nanotechnology company Zyvex and before that at Xerox's celebrated Palo Alto Research Center.[10] Gregory S. Paul, a paleontologist, and Earl Cox, an authority on the design and application of fuzzy logic systems, assert in *Beyond Humanity:* "The next century promises to be a hyper shock future, nothing like the contemporary world—nothing like what we humans are used to, or have grown to expect. All the concerns we have today, and all the plans we are making to meet them, will be swept away by the changes that are likely in the next century and those that follow—changes we have thought would take centuries, millennia, or even millions of years to come to pass."[11]

Dr. Gregory J. E. Rawlins, of the Indiana University Computer Science Department, has published two books relevant to the Singularity. *Moths to the Flame: The Seductions of Computer Technology* (1996) can even be read on-line.[12] Rawlins says bluntly: "We seem headed for our own starbirth, drawn to it just as inexorably as [interstellar] grains merge [to form stars], accelerating toward it just as surely as the merging accelerates. The attraction is massive, relentless, unstoppable. When our starbirth comes, some of us will no longer be truly human; and things we now call machines will no longer truly be machines." Rawlins is no undiscriminating booster of technology, warning: "What's special today is that because of the computer, an undifferentiated intelligence amplifier, our technology has nearly reached critical mass and is now juggernauting us around the dance floor at such a pace that we may never again be able to stop and catch our breath. Now prisoners of the dance, we're moths irresistibly attracted to the flame of technology. Prometheus, disguised as a scientist, has given us that flame. But fire also burns." Neither, though, is he a pessimist in this extraordinary moment of transition:

We, all of us, are part of the most thrilling adventure ever unleashed on planet earth. Instead of looking backward in anger and fear, let's look forward to the next dance step in the adventure we're crafting for ourselves. A century or so from now, the earth may simply be the home world of a species rich and strange, a fiercely new and amazingly interesting species—transhumanity. The human adventure is just beginning.

Let's dance.

Postcyberpunk writers such as Neal Stephenson and Bruce Sterling are developing this prospect in fiction. There is a vast, ongoing discussion among special interest groups on the Internet, extropian and other transhumanists (doughty foes of entropy; we'll get back to them), spearheaded by people like Oxford-trained philosopher Dr. Max More who, as a token of dynamic optimism, moved to California and changed his name from the less iconic Max O'Connor. They expect to be around during the Spike and genuinely hope to partake in an extraordinary upheaval—the transition, ultimately, to new forms of sentience and life, in the company of the post-Spike superintelligent machines. Dr. Ray Kurzweil, himself an extropian, has recently spread this presumption with remarkable success, in his book *The Age of Spiritual Machines* and frequent appearances in every conceivable medium, from articles and interviews in *Wired* magazine to stories in business journals such as *Technology Review* and *Business 2.0*, the Discovery Channel, and such popular American television shows as CBS's *48 Hours*.

Nor is the idea altogether new. The important mathematician Stanislaw Ulam mentioned it in his "Tribute to John von Neumann," the founding genius of the computer age, in *Bulletin of the American Mathematical Society* in 1958.[13] Another notable scientific gadfly, Dr. I. J. Good, advanced "Speculations Concerning the First Ultraintelligent Machine," in *Advances in Computers*, in 1965. Vinge himself hinted at it in a short story, "Bookworm, Run!," in 1966, as had sf writer Poul Anderson in a 1962 tale, "Kings Must Die." And in 1970, Polish polymath Stanislaw Lem, in a striking argument, put his finger directly on this almost inevitable prospect of immense discontinuity. Discussing Olaf Stapledon's magisterial 1930 novel *Last and First Men*, in which civilizations repeatedly crash and revive for two billion years before humanity is finally snuffed out in the death of the sun, he notes:

But let us keep in mind . . . another vision, in which the species' cataclysmic degeneration is not so profound . . . the ascent that follows exponentially from this premise would surpass the capacities of any artist's imagination. This means that even if the fate of humanity is not at all tragic, we are incapable of plausibly foreseeing—in the very distant future—different qualities of being, other than the tragic . . . But the existence of future generations totally transformed from ours would remain an incomprehensible puzzle for us, even if we could express it.

This is exactly Vinge's insight: that such exponentially cumulative change puts the future quite literally beyond our capacity to foresee it. The difference is that Vinge realizes how swift this change will be. It won't require humanity to await "the very distant future." But I suspect Lem, too, knew this, for he added:

It is a law of civilizational dynamics that instrumental phenomena grow at an exponential rate. Stapledon's vision owes its particular form and evenness to the fact that its author ignores this law . . . Technological development is an independent variable primarily because its pace is a correlative of the amount of information already acquired, and the phenomenon of exponential growth issues from the cross-breeding of the elements of the mass of information.

What's more, Lem understood the key factor that makes it so hard to extrapolate from what we know to what will actually come about the day after tomorrow. These innovations *interact*. You can't just alter one element of the world and leave the rest unchanged. Lem commented: "[T]he moment of the chromosome structure's discovery cannot be separated by 'long millennia' from an increase in knowledge that would permit, for example, the species to direct its development." This is a truth that stung us even before the close of the twentieth century, and now promises to utterly remake the twenty-first. From the moment the genome is finally mapped and its recipe understood, we shall begin to reshape ourselves, at first bit by bit but eventually, perhaps, entirely. What then? Lem even preempts Vinge's metaphor of an event horizon of prediction, noting "the real factors of exponential growth, which obstruct all long-range predictions; we can't see anything from the present moment beyond the horizon of the 21st century." Meanwhile, in the West, postindustrial sociologist F. M. Esfandiary, also known as FM-2030, asked *Are You a Transhuman?* (1987). And surely the most charming, preposterous pitch ever published is Professor Robert C. W. Ettinger's opening line to his 1972 book *Man into Superman:* "By working hard and saving my money, I intend to become an immortal superman."

But is it really going to happen? And should it?

Is all this enthusiasm for accelerating change, as routine and doom-savoring journalism often proclaims, just science run mad?

Is it nothing more than what Ed Regis, in his very funny book *Great Mambo Chicken and the Transhuman Condition* (1990), sardonically dubbed "fin-de-siècle hubristic mania?"

That would be a comforting and dismissive diagnosis, of course. Mockery is one way to close your eyes against that opaque, looming, onrushing future. But mockery or outrage are often the flip side of anxiety, if we're candid about it.

Plenty of doubters still confidently warn us that scientists are "playing God" or "interfering with Nature" when, let's say, a sheep is cloned, or molecular biologists manipulate somatic genes to cure cystic fibrosis (CF), or just map and file sequences for the Human Genome Project. Similar fear, one recalls, used to be directed at wild-eyed locomotive experimenters hurtling along at a breakneck twenty kilometers an hour.

Today, granted, there's far more reason to be nervous. CF remedies and cloned monkeys today, perfect teeth and enhanced IQ tomorrow, genes purchased and installed in the womb, or, trash television's favorite, the serried ranks of semihuman military clones. Still, it's quite impressive how some people have no trouble knowing precisely what God had in mind for every eventuality, including some just thought up yesterday morning in the lab.

Aside from this perpetual cry of frightened outrage from the excessively pious, many of us go into a self-inflicted cringe at the drop of a quark, or the sight of a diagram showing the coiled DNA helix like some ornate decoration from an altar in St. Peter's in Rome. Probably you'll resent me spelling this out, but look deep inside your heart and swear you've never said resentfully to yourself: "There Are Some Things the Average Man (and Woman) Are Too *Dumb* to know."

All those hideous, bristling equations. Those ferocious laws. The cascade of principles that nonscientists (I'm one too) will never start to understand in any depth, or even in any shallowness: relativity and quantum mechanics, chaos and thermodynamics, energy and entropy, theories of games and probabilities, recombinant genetics, oncogenes, T suppressor cells, melting nuclear reactors, the poisoned food chain, the crisping ozone hole, and that increasingly scary greenhouse effect.

But while we don't understand what the scientists are talking about, we certainly do know in our bones that their secret language encodes the future. That's why I've written this book: to explain how the Spike really is likely to happen, to you or your children, without going into equations, or the mysteries of the gene, or probability theory, or the mathematics of black holes . . .

Still, should we, here and now, care about the Spike? Plenty of wise men beg to doubt it. (I haven't heard any negative comments from wise women yet, perhaps because they're accustomed to thinking in generational time.) One of those wise men is Barry Owen Jones, a former minister for science in the Australian Parliament, and all-around guru of the future—*his* version of the future. Jones is a world-class polymath with honorary doctorates in science and letters, in command of vast amounts of knowledge in biography (author of *The Dictionary of World Biography*, 1994), the arts, the sciences, and industry. His book *Sleepers, Wake!* (1982, revised 1995) was among the earliest to warn of the impact of the new global economy and the technological changes underpinning it. He was a UNESCO Executive Board member in Paris and founding father of a national Commission for the Future (now defunct, alas), among the world's first official institutions to take the greenhouse effect seriously and try to work out what might be done to mitigate its impact.

His estimate of the likely immediate future is measured. In the 1970s, he notes, he "took an optimistic view of a future in which, as [Nobel physics laureate] Dennis Gabor wrote, Mozartian man (and woman) can evolve. The ending of the Cold War ended the threat of world war and nuclear confrontation, and there has been a political transformation . . . Nevertheless, population explosion, the rise of tribalism and religious fundamentalism and excessive resource use by the West suggest that we have not yet learnt to solve global problems in a rational way."[14] So Jones's response to the impending Spike, and the devastating, enthralling social impact it'll have upon us all, our children and grandchildren, is very odd. He is just not interested. "The long range future is 'unimaginable' because of the impossibility of establishing psychological engagement."[15]

We're being reminded, in effect, that nobody in Leonardo da Vinci's day needed to get in a lather about military helicopters and submarines, because they'd remain nothing more than drawings in the old genius's sketch pad for another few centuries. Why should we care about nanotechnology, when right now it's no more than a few hundred equations, a few hundred designs on a CAD screen for gears and motors at the atomic scale, a few early-model submicro devices such as microlithography pens, sensors? Mr. Jones's question is not just "why should we care"— but "how *could* we care?"

He spells this out with a sporty analogy, the kind politicians love to use when they instruct us in matters too hard for our poor little heads to take in: we can speculate about the winner of next year's football final, Mr. Jones declared near the close of the twentieth century, "or the U.S. presidential election in the year 2000—but discussing the football or politics of 2100 is too remote for serious consideration."

Is that persuasive? So much, after all, for those bothersome concerns about the greenhouse effect, which certainly won't be impacting critically upon any U.S. presidential elections in the immediate future. Even in the most dire forecasts, hothouse carbon warming isn't likely to upset the football in the developed world for at least fifty more years.

The state of the world in the year 2100 is *"too remote for serious consideration"*? Suppose it is. Let's ignore the fact that if medical and geriatric improvements continue to multiply at the current rate, some of us might be alive and kicking and in good health in another century's time. Put aside the possibility that even if the research miracles arrive too late to repair and sustain the youngest of us, our *children* will certainly stand a good chance of surviving into the year 2100.

Leave all that to one side. The oddest aspect of any easy dismissal of the looming curve of the Spike is simply this: we probably won't need to wait until the start of the next century to be swept up by its escalator. The date proposed by most of the scientists who advance the notion of an impending technological Singularity is around 2035. Less than the distance into tomorrow that's elapsed since my last day at high school, back at the dawn of the 1960s.

The Spike starts its upward curve

The Spike was first glimpsed in 1953, not by wild prophets in the desert but by American armed forces personnel working for the Air Force Office of Scientific Research. They wanted to map the path of likely change in their aircraft and missiles.

After the Second World War, all bets on possible developments in the air were off. Radar had altered warfare on the ground and in the sky, and jet propulsion was obviously going to replace the propeller-driven aircraft known since antiquity. Rockets had fallen sickeningly on London, and now their designer, Wernher von Braun, was shooting for the stars under the auspices of his former enemies.

In fact, while reaching the stars was his long-term goal, his immediate project as technical director and then chief of the U.S. Army ballistics weapon program was the delivery of heavy nuclear devices, if push came to Cold War shove, into the Soviet Union and its reluctant allies. Everything was getting faster and more powerful and more deadly; it was a clear trend. You could map it on graph paper.

Curves and trends

In fact, that's what trends are—imaginary connections drawn on suitable charts, smooth lines linking the points of inflexion of a series of S-shaped, or sigmoid, curves. Sigmoids start slow, go into a phase of rapid acceleration, then twist over and go flat again when the thing they're mapping reaches saturation. The lower half of the S is concave, the top half is convex, and the place where the convex part runs out of puff and starts to trail away is called "the point of inflexion." These curves might register rates of achievement in some enterprise. Speed, say. You can sketch the data points into a curve.

While the earliest steam car was built in Peking in 1681 by a Jesuit missionary, it wasn't until 1804 that Richard Trevithick hauled ten tons of cargo, plus threescore men and ten, at a dizzying eight kilometers an hour with his engine. A true railway train was built by George Stephenson in 1814, and his locomotive was first opened to the public in 1825. It was not, I think it's fair to say, astonishingly swift.

Things improved. By the 1840s—that is, in less than a generation—an express rail trip from London to Exeter took less than a third of the time you'd require to jolt there by stagecoach. Within another two generations, trains ran between England and Scotland at 100 kilometers per hour. In 1955, a French electric locomotive reached 330 kilometers per hour, pretty nifty, and apart from a few special "bullet trains" this kind of mad dash is seldom equaled in ordinary commercial traffic nowadays.

In effect, the curve representing travel by rail rises quite dramatically after the invention of the steam train, more or less peaks within a century, and then chugs along at a pace set by the ever more mordant costs of improving and maintaining smooth tracks, installing reliable signals, then being beaten about the head and shoulders for market share by those automobiles and the vast corporations who chose to turn from rail to road and air for their haulage. It's an S-shaped curve.

One can chart the similar rise in speed of the gasoline-powered car, from Tin Lizzies on corrugated muddy country roads to Hyundais and Porsches purring down autobahns—or, more likely, stuck for infuriating minutes at a time in gridlock. Your car might be capable of 300 kilometers an hour but you'll never see it on your speedometer except by hiring time on a specialist racetrack. The curve for aircraft starts later and rises more sharply, but its rate of climb, too, falls off with the decades. The curve always flattens again.

Flying into the future

What's happening in all these cases is simple: technical solutions are tried for the many and unsuspected problems that need to be dealt with in a new medium, using mechanisms at first balky and makeshift but quickly brought to heel and greatly improved, if never quite perfected. So your chart of modes of transport and their success stories reveals a series of graceful rising trajectories, each of them bending inevitably toward the horizontal as the costs outrun the benefits, flattening when factors such as weight and available power and air resistance and load and safety and environmental impact and a hundred other criteria and parameters dominate the mix, the equation. Within a century, or even more rapidly these days, engineering wizardry turns scientific or technical breakthroughs from working machines to accomplished elements of daily life, and everything more or less . . . stabilizes.

If you wished to get from Europe to the United States with least time wasted in an aluminium tube, you paid premium prices and flew the Concorde faster than sound—but there's no way anyone other than a fighter pilot will get there faster than that, unless hundreds of billions of dollars get pumped into a project to build a ballistic aircraft that burns all the way free of the atmosphere on rocket engines and comes down like a flying brick, as the Space Shuttle does.

But note: this sequence of independent curves makes an intriguing pattern of its own. Actually, the curves are *not* strictly independent, since they derive from the same engineering technologies that feed back into each other to some extent. Still, a train is not a car, and a car certainly isn't a plane, and you can't get to orbit or the Moon in a jet. So perhaps it is genuinely startling to find that a second-order line can be drawn through the points of inflexion of these several only loosely linked histories of transport.

That higher-order curve, then, is a *trend*, for *increase of speed as a whole*, speed available for human use, in human transportation. It starts, as we noted earlier, at walking pace and stays there for what seems like eternity. It kicks up with the taming, and husbandry, of horses and their cousins. Wind and steam and gasoline goose it again and again into overdrive, until with the arrival of spacecraft the curve is headed almost straight up to the top of the page. It's a trend—a metatrend—that looks as if the hand of God is pushing us along, with one invention peaking just as another comes along to take over the baton.

Is it a true trend? Well, yes and no. We have been moving faster and faster, just as the trend curve shows. But it can also be argued that it is—sorry—nothing more than an artifact, an illusion, a pen stroke run through a series of histories that could in principle have been connected quite differently if they'd been made in a different culture from ours. One thinks of the Chinese invention of gunpowder, used for centuries as fireworks instead of firearms.

Still, once you've drawn that curve, once you've sketched it on logarithmic paper in which each vertical interval stands for ten times the speed of the one below it . . . why, you get a shivery feeling. Perhaps the line truly is *telling* you something. Perhaps it's a kind of—Well, a hunchy predictive device.

Successful trend predictions

At any rate, that's what the U.S. Air Force guys figured back in 1953 as they charted the curves and metacurves of speed. They kept the curve running, let it press forward. It told them something preposterous. They could not believe their eyes. The curve said they could have machines that attained orbital speed . . . within four years. And they could get their payload right out of Earth's immediate gravity well just a little later. They could have satellites almost at once, the curve insinuated, and if they wished—if they wanted to spend the money, and do the research and the engineering—they could go to the Moon quite soon after that.

So the trend curve said. But, of course, trend curves are just optical illusions, created and warped by the partial, selected information you care to put into them.

Everyone in 1953 knew we could never get into space that quickly. Even the wildest optimists hoped for a lunar landing by no sooner than the year 2000 (that fabled signifier of the impossibly remote future).

The curve, however, as you know, was right on the money. Russia sent Sputnik into orbit in October 1957, and Armstrong said his little sentence on the Moon less than twelve years later. It was close to a third of a century sooner than loony space travel buffs like Arthur C. Clarke—or so conservatives had painted them, and now had to look away abashed—had expected it to occur.

Was the trend for speed actually exponential? It was starting to look like it.

Tracking the trend

Forty years ago, I learned about the trend curves that seemed to be dragging us inexorably into some kind of Spike from a popular science article by an engineer, the late G. Harry Stine.[16] Under the pen name Lee Correy, Stine also wrote rather stolid science fiction, but he made his living as an innovations promoter, managing scientific research. No laboratory drone, he saw himself as a synthesist of cutting-edge ideas and practices. He subsequently published books promoting the concept of Solar Power Satellites to beam us down cheap electricity in microwave form. His 1961 article was a deliberately provocative slap at his fellow speculative writers, usually regarded by sober citizens as lunatic technophiles. Stine denounced these specialist dreamers and extrapolators for their stick-in-the-mud conservatism.

Look at the curves! Stine cried in effect. What's wrong with you? Are you all blind? A year later, in the wonderful nonfiction book *Profiles of the Future*, Arthur C. Clarke diagnosed this same defect as Failure of Nerve, and coupled it with another crime, Failure of Imagination. Stine was determined to fall victim to neither failing. The trends were going asymptotic, he pointed out.

When you're tracking the patterns formed in the twentieth and twenty-first centuries by data on available energy, or transport speed, or numbers of people in a world afflicted with unchecked overpopulation, you can get some hair-raising and very weird results. Too weird to be true. In 1973, sociologist FM-2030 predicted faxes, satellite cell phones, and something like the Internet (good going!) together with the feasible-but-impossibly-expensive "hypersonic planes projected for the late 1980s" that would "zip you *anywhere* on the planet in less than forty minutes," not to mention the wildly extravagant hope "that by 1985 we will be able to postpone aging in a dramatic way" and onto the truly fatuous: "The use of artificial moons or satellites to control tides and floods." Curves are tricky things to interpret realistically.

"If you really understand trend curves," Stine wrote with a perfect poker face, in 1961, "you can extrapolate them into the future and discover some baffling things. The speed trend curve alone predicts that manned vehicles will be able to achieve near-infinite speeds by 1982." Perhaps that seemed safe enough back then. Two decades away. Anything could happen in twenty years. Perhaps the horse would talk. To tell the truth, Stine was concerned that this prediction might be too conservative. "It may be sooner. But the curve becomes asymptotic by 1982."

You have probably noticed that this did not happen, except on the television and movie screens when starships routinely travel at Warp Speed or burn through wormholes from one side of the galaxy to the other. It's quite disappointing. So where was the flaw in Stine's case? Surely it wasn't simply that we didn't yet know how to do such things. Science keeps learning new and astonishing things about the world. We bump into discontinuities, and have to reformulate our theories, or the theories lead us into novel facts. Sociologists of science, as everyone now knows due to the term's misappropriation by New Agers, dub these major shifts "paradigm changes."

Stine was expecting some *very big* paradigm changes.

Led astray by his transport-speed trend, he had noted: "If this is really the case, a true scientific breakthrough of major importance must be in the offing in the next twenty years." But how could such an infinitely fast vehicle be propelled? Look, say the trend curve had got a little confused, mistaking Newtonian physics for the more up-to-date Einsteinian variety. Perhaps we would settle for close to the speed of light, the best one can hope for in a universe where nothing material can go faster than light? But that costs a *lot*, it takes plenty of propellant. Pushing a hundred-ton starship up to 99.99 percent of the speed of light, as close to infinite speed as we're likely to reach any time soon, you have to pump in so much energy that the ship is a kinetic bomb carrying more than 220 million megatons locked up in its inertial mass. Energy equals mass, remember, and that's how much brute energy it takes to get a hundred tons moving that fast. Luckily, explained Stine, *that's okay!* "The trend curve for controllable energy is rising rapidly . . . By 1981, this trend curve shows that a single man will have available under his control the amount of energy *equivalent to that generated by the entire sun*" (his italics).

Oh dear. Oh dear oh dear. I don't think so. It would have taken more than a genuine "cold fusion" breakthrough to bring that one off. It would have taken a tame black hole in your tank.

If this goes on—

Stine's trend curves, in other words, were misleading. They were just what they seem to be—idle curves linking four or five quite disparate historical trajectories. Athletes and swimmers have surpassed previous records every year on the dot, as they eat better and train more cunningly and indeed come into the world as bouncing world-beaters, having enjoyed optimum medically guided conditions in the womb. It doesn't mean that some day an Olympic-class runner will smash the tenth-of-a-second mile record set an instant earlier by her trainer-sponsored rival.[17]

Does this splash of cold water mean that the Spike, the technological Singularity curving up there in the mid-twenty-first century, is nothing better than a mirage? Not at all. It wasn't sheer fantasy Stine was retailing, after all, just a rather trusting—or wickedly goading—application of the principle *"If this goes on—."*

Plenty of things are *going on* and will not stop before humankind and our world are changed forever.

Standard forecasts

In a tiptoeing sequel to his original article, published in 1986, Stine avoided any mention of his most preposterous trends and settled for merely utopian expectations:

> We can't close the Pandora's Box of technology. Technology is never forgotten; it's only replaced by better technology. Because we can't put the thermonuclear bomb, recombinant DNA, and a host of other technological wonders back into Pandora's Box and forget them, we must deal with them. It's not easy.[18]

He added: "A hundred years from now, barring an incredible combination of bad luck and poor management, people everywhere will be many, rich, and largely in control of the forces of nature. I've got faith in the capabilities of human beings. We'll make it. Therefore, we must learn how to be rich and handle abundance because we've never had to do it before."

Probably this less extravagant forecast would raise no eyebrows among comparatively conservative soothsayers. In 1996, a team from British Telecommunications made its futurological report for the period to 2020. The average Western life span by that date would be a century. Self-programming computers were expected as early as 2005, as was full voice interaction with machines. AIs emulating the human brain might exist by 2016, and by the same date genetic links to all diseases will have been mapped from the decoded DNA template so everyone will carry an individual genome record wired into a personal health card. These projections were modest, befitting a vast corporate institution like BT.[19]

Among more adventurous observers, it has been forecast that the multibillion-dollar international Human Genome Project might turn up a cure for death itself, if mortality turns out to be governed by a tractable number of genes. How likely is that? Death, after all, results from many converging factors: genetic trade-offs, oxidative stress from metabolism itself, accumulated physical damage from an abrasive outside world. Still, University of Michigan gerontologist Richard Miller was cited in 2000 as declaring that senescence, or aging, is "a single, fairly tightly controlled process that has a relatively small number of genes timing it."[20] In June 2000, the private corporation Celera and HUGO, as the project is known, jointly announced the completed initial map of the sequence of human DNA—along with the genetic instruction sets of several lesser creatures, for purposes of comparison.[21] Reading the cookbook is not the same as knowing how to make the cake, granted, but it is the crucial first step to changing the recipe.

Cynthia Kenyon, Herbert Boyer Distinguished Professor of Biochemistry and Biophysics at the University of California, San Francisco, has reported that the life span of a kind of nematode worm, *Caenorhabditis elegans,* had been increased manifold just by mutating several genes that control the creatures' rate of metabolism. This is a version of a trick they can induce themselves during lean times, putting themselves into a sluggish state of arrest known as *dauer*. The standard maturation signal gene *daf-2* is switched off during the emergency, allowing the expression of another gene, *daf-16*, that extends sleepy life span. Kenyon modified *daf-2*, creating a longer-lived, vigorous worm. *C. elegans* are simple critters, with just 959 body cells compared to the hundred trillion (10^{14}) we are made from (although those hundred trillion comprise only 254 different cell *types*), but it's an extraordinary proof of what is possible. Early in the twenty-first century we will know how to locate similar genes in humans (*daf-16* resembles mammalian *HNF3* genes), how to edit them, perhaps how to switch them on and off, or defer and moderate their influence. The worms can do it themselves just by ignoring their environment; in 1999, Kenyon and a colleague announced in *Nature* that simply depriving the creatures of genes for the senses that provide feedback from their surroundings yielded a 50 percent increase in life span. Even more remarkable, in September 2000, *Science* announced that researchers at Emory University and biopharmaceutical company Eukarion, Inc. had boosted *C. elegans* lifespan by the same amount just by adding synthetic versions of two enzymes, superoxide dismutase and catalase, which combat oxidative stress.

Is immortality around the corner?

Defeating death. Endless youth. Have we taken leave of our senses? Just so, according to former counterculture rebel Richard Neville, now an alternative futures spokesperson for the New Age. In a millennial alphabet published on January 1, 2000, he commented sarcastically (articulating the qualms of many, I suspect): "D is for Death, whose abolition by natural causes is now considered achievable, even by experts not known to be mad. Such could not be said of those actually seeking to inhabit their body forever."[22] Are we back, after all, to Stine's ridiculous straight-up-the-page curves? Or are they, too, properly decoded as a hint of the Spike, not totally ridiculous after all?

Technology will certainly remedy many intractable medical conditions (as it has begun to do), by allowing damaged DNA to be repaired or bypassed.

Cancer, for instance, paradoxically afflicts those tissues that constantly repair themselves—colon, uterus, milk ducts, the skin on both the outside and inside of our bodies. Tumors can live forever, unlike regular tissue. It is, one might say, the downside of immortality. Mastering the cellular repair system that goes haywire in carcinoma might be only a few short steps from drugs for longevity. Dr. Robert Weinberg, a notable and productive oncology researcher, argues that evolution and Darwinism (at the gene level) turn out to be the key, perhaps surprisingly.

Is cancer caused by nasty things we eat or breathe (like cigarette smoke), or by viruses, or do our tissues just lose their grip, as they do when we age? All of the above. The secret of cancer is identical with the secret of life: we are fallible but brilliantly maintained organic machines controlled by a library of genes. These are both a hoard of recipes for building proteins, and part of the factory that compiles them. If the recipes get scrambled, our cells cook up the wrong materials. To use Weinberg's own metaphor, we have genetic brakes and accelerators. Malignant cancer—uncontrolled, disordered growth—happens when the brakes (tumor suppressors) are disabled and the accelerator (growth promoters) is jammed on. Proofreading systems, and standard "suicide genes" that usually destroy corrupted cells, must also fail. Cancerous cells are so rare (given that vast number of proliferating cells in the body) because each fatal tumor requires a series of perhaps five or six random, independent proof-copying or mutational errors. The odds of all these errors striking a single cell are very small; but over a lifetime we copy cells ten thousand trillion times. The few multiple mutants to escape correction become immortal and uncontrolled, even growing their own blood supplies.

Of course, one error might make the occurrence of others a thousand times more likely (just as having a tire blow out at high speed can make your windshield more likely to shatter), but this is not part of a cancer blueprint. Nor is environmental pollution as big a problem as you'd expect. In the last seventy years, adjusting for age and cigarette use, cancer rates have *declined*. If we can persuade people to give up smoking—hard to do, at a time when the young have decided it's sexy again—cancer rates really will plummet.

What of aging itself? Can we stop senescence? The same fast-turnover tissues prone to tumors also produce large amounts of the enzyme collagenase, which destroys proteins that help to protect skin from wrinkling. Medicos might block the action of collagenase, or modify it. Even normal eating ages us more rapidly. Cutting the dietary intake of rats, mice, and monkeys by 30 percent lowers their metabolism and extends life spans by up to 40 percent. Late in 1999, as we saw above, it was announced that the life expectancy of one kind of lab mouse has been extended 30 percent *without* near-starvation, by deleting the gene *p66shc* from its genetic recipe. It is suspected that once we know just how this works, it would work with people too. If so, we already verge on knowing what we need to do in order to live much longer, healthier lives, and if lower-calorie regimes work for people (as some medical enthusiasts claim) we can start right away.

From this mix of theory and practice, of new ideas and new facts, will emerge the altered bodies and minds (human, trans-human, and otherwise) of the twenty-first century.

Editing the code

Science can already literally replace the human heart—with a baboon's, say, or an artificial pump. In coming decades, it will gain increasing power literally to rewrite the genetic code that builds each heart from protein. The fundamentals of this enigmatic new scientific and political reality are best grasped by looking at precise examples like the inherited disease cystic fibrosis. One Caucasian in 25 carries a defective CF gene, so one child in 2500 will get this awful illness. Genetic mapping already provides prospective parents a simple, inexpensive test allowing them to assess their chances of creating a damaged child. Similar techniques permit prenatal screening for various crippling disorders, so only afflicted fetuses need be terminated—which can actually reduce the overall number of abortions.

More menacing—or exhilarating—are prospects of splicing new genetic instructions into either somatic cells (bone marrow, say, in the body of someone born with a defect, an intervention that dies with the recipient and in any case needs to be topped up regularly) or germ-line cells (where the new instruction is passed onto the recipient's children). The former are now being tested in humans, while the latter are forbidden—though they are commonplace in animal experiments, when human-mouse cell hybrids have long been a useful lab tool.

Although the Genome Project will accelerate the knowledge base for such interventions, there is very much more in an organism (especially a person) than is to be found even in a total DNA map. The ethical consequences are formidable, even before the slope of the Spike turns up into its headlong overhead ascent. We need to get our thinking under way well and truly in advance. It's often said that the DNA of our cells comprises a message, written in a "genetic code" or "language of the genes." Via a dizzying chemical virtuosity, its four-letter alphabet and three-letter words construct our tissues, so each of us is "written" into existence, within a specific, rich cultural environment, from that single recipe of 100,000 genes.

One of the most intriguing and hopeful of recent discoveries concerned the role of *telomeres*, nucleotide structures that cap the ends of chromosomes and help protect their stability. Since chromosomes are strings of recipe genes and control *codons*—the design manuals, as it were, in the core of every cell—this suggested that the repair and integrity of the telomeres was itself a key to the reliable operation and preservation of tissues. That possibility is now in question, but it makes an fascinating story of how science can test new theories of longevity.

Some two decades ago, it was found that these crucial features are built out of short nucleotide sequences (*TTAGGG*) that are repeated again and again. In humans, these are strings two thousand chunks long, or should be. Electrifyingly, it appeared that normal aging was related to a design feature in standard cellular replication, apparently evolved as a precaution against runaway cancer formation. Each time a cell in your body divides to replace itself—and not all of them do so—the telomere tips tend to shorten. That doesn't happen to germ-line cells (those producing ova and spermatozoa). Apparently, special machinery was in place to guard the crucial sex cells from deterioration.

Take a human cell derived from a newborn baby, place it in a nutrient culture on a petri dish, and it will divide up to 90 times. A cell from an old person of 70 has far less kick left in it, or so it seems; it will stop replicating after 20 or 30 divisions. It has lapsed into senescence, reaching the celebrated Hayflick limit described some 30 years ago by Leonard Hayflick and associates at the Wistar Institute. Might that limit might be a by-product of telomere shortening? Perhaps cells estimated their permitted longevity by checking how much of the cap remains.

It turned out to be more complicated than this, actually, because the tips can also grow by adding on newly synthetized units. Cancer cells, as we know to our cost, regain the vivacity of youth, and it can hardly be a chance coincidence that their telomeres tend to be maintained in tip-top condition (but this, too, is not always so). Still, it seemed that if we learned how to turn off *their* telomere repair system, selectively, maybe we could defeat the tumors. And by *enhancing* telomeric repair in ordinary cells, maybe we could make them immortal.

The device that cells use to repair their chromosomal end caps is a specialized enzyme, telomerase. So we would face an agonizing choice: fight tumors by denying them telomerase (by introducing a tailored antagonist), and thereby, perhaps, hasten the body's general decline into senescence. Or enhance the longevity of all cells, while raising the risk that an opportunistic cancer will then burst into frantic, gobbling life.[23]

Telomere doubts

The case for telomerase as a key to longevity is still uncertain. In March 1999, scientists at the Dana-Farber Cancer Institute of

Harvard Medical School and Johns Hopkins School of Medicine announced in *Cell* that mice engineered not to make telomerase lost telomeres as they aged and suffered progressive defects in organs with rapid turnover of cells. Hair grayed and fell out earlier, stress impacted more severely, wounds healed slower. What's more, those cells building blood, immunity, and reproductive function significantly worsened in subsequent generations. Dr. Calvin B. Harley, the chief scientific officer at Geron Corporation, said: "This is a landmark study in telomere and telomerase biology. It underscores the potential of this field to lead to new medicines for treating various chronic, debilitating age-related diseases including cancer."

Geron's work with telomeres and telomerase, and similar pioneering studies by Drs. Woodring Wright and Jerry Shay at the University of Texas Southwestern Medical Center in Dallas, has at least partially overturned the Hayflick limit in healthy cells. Ordinary skin cells with telomerase deliberately activated have now been growing in research labs since 1997, multiplying in their sterile glass containers without error or cancer about once a day. These many hundreds of extra undamaged divisions do not mean telomerase is the key to eternal life—other factors are also involved in promoting cellular longevity, and brain cells, for example, do not renew themselves. Indeed, Cambridge University molecular biology researcher Aubrey de Grey, an authority on mitochondria and aging, has stated bluntly that "cell division is too infrequent in the body to give telomere shortening any chance of playing a role in aging." It remains to be seen whether Geron's work with the enzyme will help lengthen life span. It *is* known that Dolly, the cloned sheep, has telomeres 20 percent shorter than normal, so she and other clones might be doomed to an abbreviated existence.

One of the most tragic of all medical disorders is Werner's syndrome, which causes its young victims to "age" with shocking speed. By their late twenties they resemble shrunken geriatrics. Their arteries and heart muscles are a mess—but, oddly enough, their brains are unaffected. Werner's is caused by a defect in just a single recessive gene, of which the luckless victims possess two copies (as cystic fibrosis patients have two copies of the recessive CF gene). It codes for a helicase enzyme, which controls the unwinding of the DNA helix during replication. Loss of this key gadget stops other repair enzymes getting to the coded sequences inside each cell and "proofreading" and repairing them. The implication is that this molecular repair system might be boosted, although not simply by increasing the dosage of helicase (which, paradoxically, can be lethal).[24] Such disorders show that senescence is not a simple "natural" curse that we all must endure. Science is opening paths to improved cell maintenance and replication, simultaneously protecting against cancer and increasing longevity.

No doubt, much more detailed understanding will emerge in laboratories during the next decade or two. It is still not impossible that subtle control of telomere maintenance will help extend human life spans. After all, we know that the cells giving rise to viable sperm and eggs *are* effectively immortal. There seems no reason why the machinery that protects them should not be craftily adapted and extended to the rest of our cells. In 1996, neuroscientist and physician Dr. Michael Fossel published *Reversing Human Aging*, preaching the gospel of telomerase therapy, which he claimed would be generally available "before 2015."

The book was dismissed by a commentator in *New Scientist* magazine as "feverishly optimistic," which seems just. In view of recent papers on telomere terminals in *Science* and *Nature*, and findings by such authorities as Blackburn, Dr. Titia de Lange, a cell biologist at Rockefeller University, and Nobelist Thomas R. Cech, at the University of Colorado, Fossel has jumped the gun. Certainly, natural selection has not tumbled to this trick or, if it has ever done so briefly, the genomes of those individuals failed to pass through the evolutionary sieve. In fact, there are evolutionary pressures *causing* many species to age and die. Genotypes that emphasize efficiency in maintaining the body they build tend to leave, in the long run, fewer offspring than those making little or no provision for correcting cellular errors once the prime breeding season is done. As well, some proteins that help an infant swiftly reach healthy breeding age can have terrible, even lethal consequences later on. That doesn't matter, though, to the blind mechanism of evolution. Once your genes are replicated into offspring that will be fertile in their turn, that's the end of evolution's accounting. Yet Fossel's telomeric theory may still be correct in principle (although it will only be part of the aging story). Knowing that, we can *fix* it, *from the outside*, once we know how.

It seems very hard for people to accept this prospect, despite much media excitement over recent breakthroughs in stem-cell research and other longevity-related advances. Conservative *New York Times* commentator William Safire, in his first column for the year 2000, edged up to the possibility. "If the next-to-last year of the second millennium is remembered for anything, it will be for the discovery of the human body's ability to regenerate itself," Safire stated with surprising boldness. Citing a conversation with Dr. Guy McKhann, head of the Harvard Mahoney Neuroscience Institute, he added: "These wild-card cells may be found not just in embryos but in adult bodies, and could, in effect, reset the clock—time and again, doubling and redoubling the life span." Safire went so far as to imagine "readers of a distant tomorrow" who millennia hence will say, " 'You know, this fellow was incredibly prescient.' " Almost inevitably, Safire adds the moralistic rider for the majority of his readers who cannot readily face this outcome: "And another will respond, with all human skepticism, 'Sure, he was right—but do you really want to live forever?' "

Why not, though? Well, because of an antiquated, superstitious fear that we would thereby "break Nature's law" or "interfere with evolution's plan."

Evolution is not a planner

Despite the beautiful patterns of life, evolution *has* no plan. It is a gigantic, stupid lottery. Natural diversity, Stephen Jay Gould tells us, is usually attributable to nothing more interesting than a drunkard's walk away from a wall. Wherever she goes, the drunk will end up either smashing into the wall, or she'll topple, after a meandering course along the pavement, into the gutter. This is a dangerous metaphor, but Gould does not intend to denigrate humanity—just to rebuke our pride.

Emergent life starts simple (against the "left wall" of the complexity chart) because it can't start any other way. Mostly it *stays* simple. Even now, arguably, most of the earth's biomass is elegant, uncomplicated bacteria. Recently, archaic forms of life have been found dwelling happily deep under the crust. As much living material, simple but persistent, might be spread under our feet as floats and gallops and soars in all the familiar habitats of the globe.

Life, of course, never stays still. Mutation gnaws at each DNA message, and ruthless environmental selection winnows the alternatives. Over time, some variants grow more complex, wandering off toward the open-ended or right-hand edge of the chart. Others wander back again. Humans and other large animals exist way off on the rightmost tail of the curve. But this definitely does *not* imply that some imaginary "surging life force" has been struggling to create *us*!

Before Darwin, people supposed that God had done the design work. Evolution hinted that we could replace God-the-designer with some kind of drive toward complexity. Sadly, that notion is probably just as erroneous and self-preening. It's true that some forms have grown more complex in the billions of years since life's emergence on Earth. But this is not, Gould asserts, because there is any "complexification drive." It's a side effect of the wall over there on the left, and the vast eons of life's drunken stumbles.

In such a universe, we are freed from fears of impiety. Since evolution does not have a plan for us, we may choose one for ourselves. In fact, that is what we have always done, whether we knew it or not. So defeating death need be no more absurd a goal than finding remedies for nearsightedness, asthma—I have been on daily drugs for asthma for more than a quarter of a century, and it has improved my life beyond recognition—or, say, the lack of an ability to read and write at birth, or fly a jet by instinct.

Ultimately, we might expect to resolve the medical components of what is called "aging"—the damage and at last senescence that now accumulates with the passage of time—and find ways to outwit them. In the longest term of the history of intelligent life in the universe, it will surely prove to be the case (tragic, but blessedly brief in comparative duration) that the routine and inevitable death of conscious beings was a temporary error, quickly corrected.

Recall William Safire's assumption that human wisdom speaks against the lure of living forever. Since we have never had this option, I rather think that traditional wisdom is recommending the abandonment of impossible and bruising dreams. Once those dreams approach realization, however, the situation is reversed. Yet that attainment will not be without its inevitable strange consequences. Vinge has declared:

Radical optimism has apocalyptic endpoints—even if there are no hidden "monkey's paw" gotchas. It is interesting that the prospect of immortality leads to many of the same problems as increased intelligence. I could imagine living a thousand years, even ten thousand. But to live a million, a billion? For that, I would have to become something greater, ultimately something much, much greater.[25]

A clone in sheep's clothing

Perhaps the most startling biological breakthrough of the end of the twentieth century was Dolly, the cloned sheep. Oddly enough, everyone had been awaiting the arrival of cloning, and yet nobody expected to see it—especially the scientists, who are usually the most conservative when it comes to predicting the near future.

Dolly was already seven months old by the time *Nature* published details of her unorthodox conception. Dr. Ian Wilmut's team, at Roslin Institute in Scotland, took the nucleus of an adult udder cell, tweaked it in various ways, and transplanted it into an unfertilized ovum purged of its own DNA, then implanted the viable embryo in a surrogate mother sheep. The task was not easy. Many hundreds of attempts were made before the successful pregnancy. Still, now that the method has been proved, the technology of cloning is swiftly maturing. We already know that clones made by nuclear transfer into eggs from donors—the method used to build Dolly—works quite well with a variety of species, but individual cloned "twins" can turn out quite different from each other! In December 1999, Roslin Institute scientist Keith Campbell, who worked with Wilmut in creating Dolly, announced that ram clones had diverged with age. "You would not know they are clones," he said, since they now vary in size, appearance, and temperament. Why so? Perhaps mitochondrial DNA and other factors in the donor eggs' cytoplasm trigger the nuclear DNA in different ways.

By the time any experiments are made in human cloning— banned in many countries, but sure to be attempted sooner or later—the procedures will certainly need to be fully understood and reliable. That, after all, is the way it worked with *in vitro* human fertilization, commonplace today despite initial skepticism and furious ethical debate. For now, ethicists remain deeply concerned by the cloning prospect. Brave New World! A series of inevitable, but dubious, horror stories have been presented in the press and on television:

Saddam Hussein or Adolf Hitler cloned into ranks of storm troopers. Hardly likely. Even if character is genetically ordained, which is doubtful, why would they obey their sarge? Besides, armies, like any organization, need diversity and variation (however much they appear to strive to stamp it out).

Rich old men purchasing identical heirs, bypassing nature's plan. This is possible, but interfering with an imaginary plan—"playing God," as it's called—is not what's wrong with the idea. As we've seen, nature doesn't have a "plan." Nature, as Darwin showed us a century and a half back, is a blind, heedless machine that eats its children and kills the "unfit." Indeed, it kills the "fit" as well, soon enough. We can do *better* than nature— let's hope so, anyway.

It goes against God's express prohibition. Oh? Which chapter

and verse in the Bible or the Koran forbids cloning by the trans-
plantation of adult DNA into an enucleated, unfertilized egg? It
is true that the Pope very quickly denounced the practice, declaring it sinful. He said the same about *in vitro* procedures. It
was the same Pope, one recalls, who only recently apologized to
the memory of Galileo and Darwin, admitting that they had been
treated badly by the Church, which for decades if not centuries
had denounced their teachings as wicked. Perhaps in a few more
decades or centuries, cloning will be pardoned as well.

Cloning steals women's sacred reproductive powers. Well, not just
yet. If you wish to clone yourself—illegally—a human womb will
still be needed, and a woman's ovum, which calls for cooperation
rather than theft. In the medium term, it's true, there might be
gene-engineered animal wombs. In the long term, a synthetic
uterus. But it's not necessarily a matter of sexism. Many women
will wish to use these services.

Actually, much of the uproar has been absurd from the start,
based on a faulty understanding of how cloning operates. We do
not go into metaphysical hysteria when twins or triplets are born.
Yet any group of genetically identical humans, whether created
by design or "natural" accident, is essentially just that. Are we
terrified that quintuplets "share a soul," or must tussle for one?

As for fascistic breeding programs—do we see them right now?
No. Are there well-funded orphanages filled with teams of children produced from the *in vitro* fusing of spermatozoa and ova
from the powerful, the brilliant, the athletic? None that I've heard
of. It could be done. It could have been going on for centuries,
millennia. Why should cloning alter our reluctance to breed babies like sheep?

Actually, cloning makes *less* sense than a harem, if you're planning to make the world over in your own image. After all, your
cloned copy isn't exactly the same as your twin, even aside from
the fact that we hardly ever find one identical twin thirty years
younger than the other.

As we've seen, somatic cells get old and tired. When they are
copied in the course of life, errors creep in. So cloning yourself
from a bit of your own tissue means the new baby begins with

damaged DNA—not a terrific start in life. Not to mention the possible impact of telomere degradation in adult cells. Poor little Dolly is some years closer to the Hayflick limit than her woolly playmates. It is now known that her telomeres are a fifth shorter than usual, so perhaps she will keel over earlier. Not that this matters with lamb chops on the hoof, but it certainly does with human babies. So if anyone wished to adopt their own healthy cloned copies—bearing in mind that they might *not* turn out identical anyway—it might be advisable to retain frozen samples of their own embryonic tissue, not something that hospitals and labs do just yet. At earliest, this would be an option of the children born into the twenty-first century. On the other hand, it is now known that stem cells from your own body can be provoked into growing any kind of tissue required, and so you might imagine growing an entire backup body—perhaps without a functioning, aware brain—from your own stem cells. As we shall see, this is surely not the way to go; it is a misunderstanding of the technological possibilities, let alone the moral issues. But we need to think this through.

It is usually said flatly that these suggestions are simply morally outrageous, not to be considered for a moment. Mightn't rich dictators have duplicate, younger bodies grown and exercised, in cloning farms, for use as hosts in brain transplant rejuvenation schemes (a standard horror scenario, despite the horrendous technical obstacles to brain transplants)? When their bodies are worn out, just call in the neurosurgeons and have their brains popped into a fresh new body. This scenario is remotely possible, but it misses two fairly obvious objections (aside from the fact that we already have laws against slavery and mutilation).

First, your brain is as old as the rest of you. If your heart and liver and eyes are wearing out, your brain probably hasn't got much longer to go either. In fact, many people deteriorate and die precisely because that incredibly complex and vulnerable organ, the human brain, has failed ahead of the rest of the organs. Besides, I wouldn't wish my brain to be grafted into a pre-aged clone.

Second, it is extraordinarily tricky to repair severed nerves, even in fingers or accidentally amputated limbs. Can you imagine what's involved in the wildly difficult task of disconnecting a brain from its sensory organs, and the rest of its body, and re-wiring it into another body? The excruciating pain as the nerves learn their new connections? The tedium and frustration of learning every basic skill again like a baby—but a baby with an adult mind trapped inside the skull.

In the long run, using some of the future technologies to be explored in later chapters, even this might be feasible without discomfort. Tiny machines no larger than molecules might pour into the brain through the bloodstream, nipping and tucking and tagging and rewiring. But the point to keep in mind is that once we have *that* level of technology, with the artificial intelligence support systems needed to run it, we won't *need* anything as coarse and morally offensive as transplants into cloned duplicates.

Besides, to repeat: a cloned double is your *twin*, and a young, defenseless child at that. If you had the chance today, would you treat your own twin as nothing more than a convenient assemblage of spare parts? Injured in a terrible accident, would you happily order your twin's brain removed so your own could be implanted in the healthy body? I didn't think so.

Could one attempt to sidestep the implications of atrocity by growing a replica body with the genetic pathway for brain development switched off? Thus, a brainless cranium, presumably "soulless," might await its new tenant.

Not a good idea. Aside from ethical repugnance, natural genetic or developmental errors such as microcephaly (tiny brain), or, worse, anencephaly (no brain) result in deformity of the body. In the anencephalic case, such damaged babies die shortly after birth. The complete brain isn't just the organ of thought and feeling—it's the control center for the entire body's development.

So the impact of the cloning breakthrough will be more modest than alarmists fear. We will see benefits from the science it yields, now that specialists can test their theories using experimental and control animals that are literally identical (aside from inherited factors from those parts of the cell not in the nucleus, such as maternal mitochondria—which might not be insignificant). One pathway sure to yield benefits is selective growth of compatible organs grown from your own stem cells. In 2000,

Japanese researchers led by Tokyo University biologist Makoto Asashima, announced that they had grown frog eyes and ears simply by cultivating embryo cells in different concentrations of retinoic acid, which somehow triggers the expression of genes needed for the different organs. Kidneys developed by a similar process had been transplanted into frogs, which had survived longer than a month.[26] In the long run, such methods will teach us many of the answers we need to know to protect both body and brain against damage, deterioration, and perhaps death itself.

Waiting in the freezer

Yet even if it turns out that nothing will stay the Grim Reaper this side of the Spike, all is not lost. An answer has been suggested—cryopreservation, or very deep freezing—that could harbor you into the borderlands and beyond. The hope is that the medical science of the next few decades (or even centuries), surely bound to be more advanced than ours, will finally gain the know-how to revive you from temporary death.

Forty-odd years ago, that method was nothing better than a narrative device in fanciful stories. When Dr. Robert C. W. Ettinger suggested it seriously in his 1964 book *The Prospect of Immortality*, few took him seriously.[27] Today a number of private companies exist that will accept your money or insurance policy and contract to preserve your corpse (as your temporarily defunct person will be crassly regarded by many, including the authorities) in a chilly storage medium for as long as it takes.

Meanwhile, cryonicists are pressing ahead with revival research, within the sadly restricted limits of their budgets. Ultimately, the cryonics supporters hope, your preserved, undeteriorated, but inanimate body and brain will be thawed, its damaged condition made good (which calls for repairs first to whatever killed you, and then to any further injuries inflicted postmortem by the cooling and thawing protocols), and you will awaken. Ralph Merkle, who expects the repairs to be made by nanomachines, put it like this in a review paper:

Cryonic suspension is a method of stabilizing the condition of someone who is terminally ill so that they can be transported to the medical care facilities that will be available in the late 21st or 22nd century . . . While there is no particular reason to believe that a cure for freezing damage would violate any laws of physics (or is otherwise obviously infeasible), it is likely that the damage done by freezing is beyond the self-repair and recovery capabilities of the tissue itself. This does not imply that the damage cannot be repaired, only that significant elements of the repair process would have to be provided from an external source.[28]

The danger that the freezing process itself damages the tissues to such an extent they cannot be repaired in the future has always been recognized. Originally, in the 1960s and 1970s, whole bodies were drained of blood and perfused with a chemical cocktail designed to prevent ice crystals forming inside the cooling tissues, crystals that would lacerate cells hideously from within during rewarming. Today, superior perfusants are under development, and many clients choose the much cheaper method of having just their heads frozen—"neurological suspension," as it's politely called. The reasoning, however macabre, is that by the time successful thawing is in place, superior technology should have no trouble cloning a new, youthful torso to replace the sacrificed portions.[29] How much fun you'd have acclimatizing to a world at least several generations sundered from your own, and perhaps already transformed by the convulsions of the Spike, is less often canvassed.

Two obvious answers spring to mind. We may hope that counseling and psychological practices will improve in line with the requisite medical advances. Indeed, by the time cryonics clients are ready to be awakened, perhaps technology will provide neural chips and enhancers to bring the revived dead swiftly up to speed. The second answer is grimmer: if you don't like the brave new world, you can always . . . well . . . *kill yourself.* Permanently, this time.

In the meantime, not too many are opting for cryo services,

sometimes not even those who have signed up. The celebrated
guru and transhumanist Timothy Leary, who died in 1996, declined at the last minute to confirm the procedure, although a
crew was on standby. However, Ettinger's late wife Mae was cryo-
suspended in March 2000, as was futurist FM-2030 in July 2000,
and psychology professor Dr. James Bedford has been in suspension, moved from one cryonics support organization to another,
for more than three decades. Bedford died of renal cancer aged
73 on January 12, 1967. Volunteers injected his corpse with protective fluids and slowly lowered his temperature in a liquid nitrogen bath to minus-196 degrees Celsius. If the Spike is on
schedule, he might have to wait as long again, or more, for resurrection.

Some 84 cryonics patients are now preserved in liquid nitrogen
at four different cryonics companies in the United States.[30] As
Ettinger put it many years ago, with a certain bitter whimsy: *many
are cold but few are frozen.*

Personally, I'd prefer to avoid cryonic methods by outliving
the need for them, so I'm also eager to see metabolic and genetically engineered repair processes retrofitted when those techniques come on line. I want my damaged teeth fixed, and my
hair back—and my *back* back, for that matter. And, yes, vigorous,
indefinite longevity would be a useful bonus. I wouldn't sneer at
physical immortality.

As we shall now see, such hopes for Promethean technology
are very far from being just a wistful fantasy. The Spike could
change everything utterly, in ways too ruinous and horrifying to
regard with merely human gaze. But then again, it might bring
a kind of . . . *transhuman redemption.*

2: Most Everything for Free

A THRESHOLD EVENT WILL TAKE PLACE EARLY IN THE 21ST CENTURY: THE EMERGENCE OF MACHINES MORE INTELLIGENT THAN THEIR CREATORS. BY 2019, A $1,000 COMPUTER WILL MATCH THE PROCESSING POWER OF THE HUMAN BRAIN—ABOUT 20 MILLION BILLION CALCULATIONS/SEC. ORGANIZING THESE RESOURCES—THE "SOFTWARE" OF INTELLIGENCE—WILL TAKE US TO 2029, BY WHICH TIME YOUR AVERAGE PERSONAL COMPUTER WILL BE EQUIVALENT TO A THOUSAND HUMAN BRAINS. ONCE A COMPUTER ACHIEVES A LEVEL OF INTELLIGENCE COMPARABLE TO HUMAN INTELLIGENCE, IT WILL NECESSARILY SOAR PAST IT . . . THE NEXT 20 YEARS WILL SEE FAR MORE CHANGE THAN THE PREVIOUS HUNDRED. THE KEY TO AN ASSESSMENT OF FUTURE TRENDS IS TIMING, DETERMINING HOW MUCH PROGRESS CAN REALISTICALLY BE EXPECTED IN PARTICULAR TIME FRAMES.
—RAY KURZWEIL, *R&D MAGAZINE, 1999*

Aside from the astounding biological changes considered in the previous chapter, a number of technological advances will have immense impact on our lives. In increasing order of importance, these include full-scale, sensory immersion virtual reality (VR); molecular manufacturing (also known as nanotechnology); and genuine artificial or machine-augmented minds (AI) leading swiftly to superintelligences (SI).

Virtual reality

VR might seem to be a highly ballyhooed technology that has already outrun its promise, but that would be to mistake the hype for the developing reality. Year by year, people in the technically advanced world spend ever more time escaping the stresses or boredom of ordinary life by holidaying in media fantasy: movies, television, computer games, and pornography, the net's virtual, on-line community itself. VR will be an important component of this ultimate "bread & circuses," since it places you inside the entertainment of your choice (to some degree) and allows you to interact with it, as if you were actually there. Aside from entertainment, VR has plenty of uses in medicine—even surgery— and research into areas where eyes and hands just can't reach by themselves.

VR is already available in modest incarnations in amusement arcades, or even running on your home Pentium, if you invest in a data glove and stereo goggles. It's the designer reality of William Gibson's *cyberspace,* the interactive computer realm of constructed data worlds, the next step beyond monitor, mouse, and spreadsheet—but what a step, coupling your human sensorium with the computer's running program.[1] VR is to the program's simulated universe what feeling and action are to the thinking mind inside the skull's shell: data bits rendered into sound, sight, and touch. Push, and the computed cyberspace world pushes back.

That can make it easy to wander through a somewhat simplified simulation of your new house or workplace, built in an architectural design program's imaginary space as you see its rooms and to some extent feel yourself walk through them. You can stroll—or "fly"—through the great art galleries and museums of the world, although the works on display will not have the rich fractal density of material reality. More usefully, chemists can reach into an atomic model of a protein, twist it this way and that, bring another molecule alongside and play with the two until they fit together. Such virtual Lego is no toy, since what the computer program is doing in these displays is to calculate, using the best available physics and laws of chemistry, just how such molecule-scale structures must affect each other. That estimate could be run in a coldly abstract way, and the results presented in a chart, but many scientists are starting to develop an up close and personal feel for this deep realm that can't be gained any other way, unless you happen to be gifted with the sublime visualizing skills of a scientific genius.

More worthwhile still is the emerging fusion of VR systems, magnetic resonance scanners, and robotic extensions of a surgeon's fingers. MRI and other subtle scanners create an image of the invisible, inaccessible insides of a patient's body. Instead of showing these pictures on a monitor, it is now possible to display stereo images in light headsets—electronic spectacles. The image can replace direct sight, or be superimposed over what the surgeon's eyes detect directly. Robotic cardiac bypass technology made it to the American Heart Association's top ten list of the best medical advances in 1999. "Three holes in the chest allow the insertion of a light/camera unit and two instruments, each held by separate robotic arms. Sitting at a console, the surgeon maneuvers handles shaped like microsurgical instruments that are connected to a computer. The computer controls the robotic-held instruments, which duplicate the surgeon's motions."[2] These robotic intermediaries can help compensate for the inevitable tiny vibrations that affect even the calmest, most rock-steady medico's hands.[3] And knowledge can be gained even in the absence of real patients, useful for training students (VR "cadavers," or devices such as the Boston Dynamics open surgery anastomosis trainer, and the drolly named EVL eye, from the University of Illinois's Electronic Visualization Lab)[4] and for pure research: a Rensselaer Polytechnic Institute nuclear engineering scientist, Xie George Xu, has designed a three-dimensional "virtual man" able to mimic accurately the damaging impact of radiation on skin, eyes, bone marrow, and gastrointestinal tract.[5]

The addictions of VR

If virtual reality enables you to enter cyberspace renderings of the brain and body, or more appealingly the great art galleries, and even the world's torrential financial transactions, gliding through n-dimensional embodiments of these vast data flows, won't the handicaps of ordinary flesh grow irksome? Might we not start to lose our attachment to the body?

"Reality," stated graphics guru Alvy Ray Smith, talking about the maximum rate of image components the human eye can deal with, "is 80 million polygons per second."[6] (Flies get by with just 10,000 picture elements, or pixels, and trained humans can read or move through a maze with as few as 625.) We are now only at the stage of holography—laser-created three-dimensional imagery—and military equipment that beams information into your eyes using a special visor, but there's little doubt among enthusiasts that Aldous Huxley's *Brave New World* "feelies," first predicted in 1932, are finally on their way. ("There's a love scene on a bear rug; they say it's marvelous. Every hair of the bear reproduced.") Yet isn't virtuoso virtual reality, like the much-hyped cascade of computer power, just the usual wretched consumer hype in post-Nintendo garb? The imaginary schtick of special effects extravaganzas like *The Matrix*?

No, it's not. While we are only just properly launched on the upward curve, available technology already exceeds most people's sense of what's possible. Clad in a sort of wired wet suit, with stereo visual and auditory headset, you can dance with a four-meter purple lobster, run through tall grass as a lion, or transfer your point of view into a robot body on the far side of the room that gazes back at your own distant flesh. The experience of disembodiment is so real, as your data-gloved hands control the robot's movements and your gaze shifts its lens, that it can give you psychic whiplash. Uncoupling from the VR rig can cause "simulator sickness." Inside the detailed factual engineering reports of VR entrepreneurs lurks a disruptive demon. The silly New Ager slogan suddenly has genuine force: in cyberspace, you *do* "create your own reality."

A range of uses has already been tested. Emergency rescue teams train in VR simulations to combat terrorist attacks that release a biological weapon.[7] Researchers at the Georgia Institute of Technology have created an on-line "identity game" to allow virtual community designers and members to explore varieties of the self they project.[8] David Friedman, professor of law at Santa Clara

University, foresees this medium, long before it becomes the total-immersion, full-presence "holodeck" of *Star Trek,* becoming pervasive in the workplace:

> The year is 2010. From the viewpoint of an observer, I am alone in my office, wearing goggles and earphones. From my viewpoint I am at a table in a conference room with a dozen other people. The other people are real—seated in offices scattered around the world ... It is sufficiently real for the purposes of a large fraction of human interactions—consulting, teaching, meeting. There is little point to shuttling people around the world when you can achieve the same effect by shuttling electrical signals instead. As wide band networks and sufficiently powerful computers become generally available, a large part of our communication will shift to cyberspace.[9]

VR's worst danger was foreseen by Arthur C. Clarke in 1949 in a description of a far future city sedated by the ultimate media: "The greater part of the city was no more than a vast honeycomb of chambers in which thousands could dream away their lives. This artificial world was utterly real to the beholder."[10] Full-density VR will be both more seductive and perhaps less soporific than this nasty vision.

But is such a prospect actually all that horrifying? If a simulated reality is equal in texture to the world of ordinary experience, yet richer in its range of possibilities (you can redesign the basic laws of physics in there, after all, not to mention your own appearance), and if its inhabitants remain in contact, able to share their joys and dreams and sorrows and plans, would you deny yourself the experience? A quarter century ago I was sure that most people would. Now I'm not at all sure—and not even sure that they *should* do so (on the suspect basis of "authenticity"), given the opportunity. Australian writer Greg Egan has made an explosive impact in the last decade or so by projecting just such scenarios in prodigiously inventive and scrupulously detailed "thought-experiment" novels such as *Permutation City* (1994)

and *Diaspora* (1997). These fictions take it utterly for granted—
and draw the startled reader into complicity with this opinion—
that such a merger of human minds and computer-mediated or
generated "realities" is not only inevitable, but far superior to our
current restricted existence.

The ultimate outcome for such permeable, permutated beings,
sharing a common contrived reality only partly interpenetrating
the empirical world of spacetime physics, has been diagnosed as
Metaman by biophysics polymath Dr. Gregory Stock. Director of
the Program on Science, Technology, and Society at the Center
for the Study of Evolution and the Origin of Life at the University
of California, Los Angeles, Stock says Metaman starts as "that
part of humanity, its creations and activities that are interdepen-
dent—joined together by trade, communication and travel . . .
Metaman is presently crystallizing out of the totality of human
endeavor that has been building and deepening for millennia
throughout the world . . . Already, the time of routine visual tel-
ecommunication, clean and cheap power, bio-engineered plants
and animals, computer-synthesized realities, and even human-
machine hybrids is beginning . . . Metaman affirms that we are all
connected—giving to and drawing from one another as we par-
ticipate in a momentous step in the evolution of life."[11]

What this vista omits or underestimates is the growing worldwide resistance to such important novelties as genetically engineered food crops and animals, and to Promethean technology
in general to the extent that it directly affects our ordinary lives.
In 1990, Stock notes, failing to foresee the headlines of Green
panic less than a decade later, "Germany enacted a comprehensive
and restrictive law to regulate research in genetic engineering and
is now finding that the result has been merely to reduce German
competitiveness in this important field." At the start of the
twenty-first century, it begins to seem that this analysis misfired,
not grasping the fearfulness with which people would greet
"Frankenfoods" and related technologies, and the zeal with which
political representatives would buckle to largely ungrounded anxieties.

My local council recently banned genetically modified food.

Reporting this, the local city newspaper opened its story with a shuddery, gratuitous mention of *Soylent Green,* the 1973 movie about an overpopulated future dependent on "synthetic" food. (Soylent Green, you might recall, turns out, revoltingly, to be people.) "No one," the journalist went on blithely, "is suggesting that the newly emerging genetically modified food industry is grinding up human corpses, but . . . there are still too many unknowns about the effects of biotechnology to be silent on the issue."

It seemed to me a disgraceful leap, disturbingly reminiscent of racist jibes that some ethnic group or other murders and eats the babies of their persecutors. *No one is suggesting,* says the reporter disingenuously, having done just that. Such is the odium science will increasingly attract early in the twenty-first century.

Granted, it's prudent to pause before introducing proprietary crops gene-engineered for high resistance to pesticides. Widespread adoption could leave us eating food with more pesticide residues, make farmers dependent on modified seed, a kind of agribiz monopoly. Monsanto famously abandoned its sterile "terminator gene"-engineered seed due to public outrage and a shareholder campaign, which saw this mechanism as a ploy forcing farmers (many of them poor Third Worlders) to buy new seed stock every year. Rather inconsistently, this criticism is often heard side by side with scary tales of runaway genetically modified pollen rampaging and crossbreeding through neighboring fields—something the sterile seeds specifically *can't* do. So these are serious issues. But to cite *Soylent Green* in that context is reckless escalation of the rhetorical stakes, instant overkill, yet increasingly typical of the standoff: demonization of useful science and technology—as well as, more appropriately, of dubious global multinational policies—driven by fright, ignorance, malice, or political opportunism.

That's not to say science is above reproach, off-limits to criticism. On the contrary, its palpable power and ubiquity demand that we keep the strictest eye on scientific research and its (often state-funded) practitioners. Instead, what we find increasingly is a slide from sensible caution to jokey but grotesque anxieties, borrowed literally without a moment's thought from the lucrative nightmares of Hollywood hacks. A former Roslin Institute researcher, where Dolly the cloned sheep was created, has cloned sixteen Friesian heifers in New Zealand after inserting the human myelin basic protein (MBP) gene which grows myelin around nerves. David Wells plans a herd of 1500 genetically modified cows. Extracted from the milk, MBP might help multiple sclerosis sufferers. Another company, PPL Therapeutics, intends to build up a flock of 10,000 transgenic sheep with the human gene for alpha-1-anti-trypsin, also extractable from their milk and used to help victims of cystic fibrosis. Green protest groups such as Revolt Against Genetic Engineering (RAGE) have denounced these plans as "Frankenstein's farms."[12]

One expert prognosticator who favors the bioengineered path is Freeman Dyson. A very practical visionary, Dyson is one of those extraordinary scientists who speak to us directly, in the tradition of Jacob Bronowski (*The Ascent of Man*) and Carl Sagan (*Cosmos*). An applied mathematician, he is more distinguished than either, yet his name is not as widely known. It should be. Dyson worked on military systems during the Second World War. Later, he helped develop a highly stable, safe nuclear reactor for making medical isotopes. In some ways he seems the very image of the mad scientist: he designed an extravagant space propulsion system using a stream of small atom bombs to blast the craft forward. Yet his book *Weapons and Hope* helped bring sense into the nuclear-weapon standoff of the 1980s.

Dyson blends the grandest long-term visions with a poignant humanism. Recently, he has helped devise an improved method for detecting land mines. He proposed that stellar civilizations might most efficiently mine the free thermonuclear energy of their suns by converting all their planetary mass into orbiting habitable artificial shells, since dubbed "Dyson spheres" or (more accurately, since they would not be solid but more like well-behaved asteroid belts) "Dyson shells." At the farthest extremes, he projects the status of life in an ever-expanding cosmos, arguing that intelligence can persist forever even in a vastly frigid open universe. Lately he has turned his gaze on the immediate future.

It is, inevitably, a world deeply influenced by new scientific knowledge and technologies derived from those insights. Marking the radical change he and his colleagues have wrought, Dyson no longer sees space as an arena for such Promethean feats as bomb-powered flight. Instead, he extols small, smart automated probes, the kind that crawl around Mars and send us back breathtaking images of worlds inhospitable to humans.

Indeed, Dyson's current emphasis is on the small world, not the cosmos. Unpacking the message of the human genome, plus the genetic instructions of many other living things, will re-make the future in surprising directions. The Human Genome Project's acceleration is due to the tremendous acceleration in computers, and to the communication revolution they underpin. Those with access to the Internet are in almost instant direct contact, at very little expense. New improvements in solar energy technology will also break down the power and oppression of great central authorities, the kind needed in this century to pro-vide electric power (and terrifying military might) to industrial societies. That won't just have its impact on us, the increasingly postindustrial world. Cheap available power can rebuild tradi-tional communities as well, saving them from destitution and abandonment by their brightest children.

Dyson's vision is grounded in specifics. His wife grew up in an East German village ruined by the collapse of communism, but now recovering due, ironically, to middle-class nostalgic gen-trification. "How can a godforsaken Mexican village become a source of wealth?" Freeman Dyson asks. Sunlight, he notes, is not restricted to the rich countries—quite the reverse. Genetic engi-neering of new kinds of crops will make solar energy available even to the poor. "An energy crop could be a permanent forest of trees that convert sunlight to liquid fuel and deliver the fuel directly through their roots to a network of underground pipe-lines."[13] It is a vision eerily similar to one nanotechnology scheme we shall examine in chapter 4. Finally, the Internet will connect village and larger world, bringing in cheaply all the information needed to develop and even employ local talent.

Oddly enough, Dyson is not terribly taken by Vinge's notion

of a technological singularity. In a *Wired* interview with Stewart Brand, he stated: "The technical tricks these people are talking about are only a small part of the human experience . . . It's true that the price per megaflop is going down according to Moore's law, but what you can do with the processing power isn't increasing at the same rate."[14]

Nanotechnology

Meanwhile, though, the "real world" will not remain very much like the limited landscape we inhabit today, heir of the biblical injunction that humans should earn their bread in the sweat of their brow, preferably with much salutary suffering to stiffen their moral spines. We are about to enter an age where production is not just automated—that somewhat delayed ambition of the mid-twentieth century—but turned over to tiny factories the size of viruses, perhaps able to replicate themselves at no cost to their users and beneficiaries. Some of this will be done by clever bioengineering, but more will be due to devices smaller than dust. Dyson disdains this view, declaring: "Biotechnology has moved ahead so fast that it makes nanotechnology old hat. If we get to the point of building micromachines, it will probably be done by biotech." With due deference to Dyson's wisdom, there are reasons to doubt this. At most, the world as it approaches the Spike will blend these two technologies of the miniature.

Molecular nanotechnology was conceived a quarter century ago by Eric Drexler, who now promotes it through the Foresight Institute in California. For his pains, he was long dismissed and mocked as a visionary crackpot. Slowly, however, nanotechnology has become a formidable and growing field of research in reputable laboratories. It is known breezily to its familiars as "nano" or "nanotech," and to me as "minting," as I mentioned in the previous chapter. The projected technology proposes machines and tools the size of molecules, a few hundred or thousand atoms in size, ideally self-replicating, able to follow instructions and ultimately build anything your heart desires out of . . . well, again ideally, out of garbage.

More precisely, in 1996 Dr. Ralph Merkle defined it thus: "a manufacturing technology able to inexpensively fabricate, with molecular precision, most structures consistent with physical law." He added cautiously that in consequence "nanotechnology is today theoretical. We do not yet have the defined ability, and it is unlikely that we will develop it in the near term (i.e., the next five to ten years)."

Until the middle of the twentieth century, the idea that we might *ever* be able to manipulate individual *atoms* seemed impracticable, if not preposterous. Then, in a famous 1959 speech, Nobel laureate Richard Feynman lent his prodigious authority to the notion, remarking, "The principles of physics, as far as I can see, do not speak against the possibility of maneuvering things atom by atom. It is not an attempt to violate any laws; it is something, in principle, that can be done; but in practice, it has not been done because we are too big."[15]

That is no longer true, even if moving atoms one by one still can only be done with plenty of patience and usually with expensive equipment. In 1996, materials scientists at Nottingham University, in the U.K., announced that they had controlled the motion of a single carbon-60 molecule *at room temperature*. Using a scanning tunneling microscope (or STM) that prods at the electric fields of the atoms under its "gaze," Dr. Peter Breton and his colleagues steered this buckyball (more formally, a carbon molecule of buckminsterfullerene, or fullerene) although it was only a billionth of a meter in diameter. Shunt buckyballs around with your STM—machines that are no longer rare and arcane; one of them has already been built as a high-school project—and you can start to compile supremely exotic novelties: wires a molecule thick, cheap synthetic diamonds, radically new polymers.

Of course, you can only produce them in tiny amounts with a STM. Scaling up to industrial productive levels might be quite a chore. Still, once the existence proof is there on the table, people with money to invest start to look with greater respect. So with

Breton's achievement the dream of Eric Drexler and Ralph Merkle was one step closer to realization.

Machines with parts built to operate at the nanoscale have now been available for some years, although they do not quite fit the bill. Enormous excitement was generated in 1997 when an Australian team led by Dr. Bruce Cornell, in a research partnership between the Commonwealth Scientific and Industrial Research Organization (CSIRO) and industrial consortium AMBRI Ltd., announced a "nanosniffer" or biosensor with an ion-channel switch just one and a half nanometers across. In a way, this nifty gadget reproduced the chemical sensitivity common in living creatures, and indeed its design was borrowed from biology. Its synthetic membrane, which is bonded to a thin metal film, allows different ions to pass through selectively—and more cheaply and reliably than previous methods. The sensor, which can in principle detect a sugar cube dissolved in a harborful of salt water, will pick out viruses, drugs, pesticides, even specified gene sequences. The project has proved so successful that the need for government funding, critical to its initial development, has now been terminated. Meanwhile, chemistry professor Michael Wilson, at Sydney's University of Technology, heads several labs working on a nanoscale motor, with working parts in the billionths of a meter.

A still more remarkable step was announced in a December 1999 issue of *Science:* Philip Kim and Charles M. Lieber, of Harvard, have constructed what they dub "nanotube nanotweezers." They are able to catch up and manipulate tiny clusters of silicon carbide molecules and gallium arsenide nanowires. While these tweezers might still be too blunt to handle individual atoms, they could be forerunners of gadgets able to move and stack molecular building blocks, enabling the construction of very small devices—just the kind of machines Drexler and his colleagues mapped out several decades ago. And meanwhile, new properties of fullerene nanotubes suggest dizzying levels of miniaturization. At the close of 1999, labs at the University of Pennsylvania and elsewhere reported on ways to create diodes using nanotubes—the crucial component of electronic switches, and perhaps the basis of a major and unexpected jump downward in size and cost of computing.

As if this were not sufficient proof of the real-world viability of such concepts, a November 1999 issue of *Science* reported that Cornell University researchers H. J. Lee and W. Ho had bonded a carbon monoxide molecule to an iron atom (at a chilly 13 degrees Kelvin, it's true). So compiling interesting molecular structures in a controlled manner was no longer theoretical, let alone wishful thinking—it has been achieved, if only in a small way (so to speak).

Still, until recently few commentators gave much credence to molecular manufacturing. Build things from the atom up, rather than carving bulk materials down to the scale you needed? It seemed absurdly hubristic. Actually it was nothing of the sort. Dr. Roald Hoffmann, winner of the 1981 Nobel prize in chemistry and on the technical advisory board of Molecular Manufacturing Enterprises, has commented: "What is exciting about modern nanotechnology is (a) the marriage of chemical synthetic talent with a *direction* provided by 'device-driven' ingenuity coming from engineering, and (b) a certain kind of *courage* provided by those incentives, to make arrays of atoms and molecules that ordinary, no, extraordinary chemists just wouldn't have thought of trying. Now they're pushed to do so. And of course they will. They can do anything. Nanotechnology . . . is the way of the future, a way of precise, controlled building, with, incidentally, environmental benignness built in by design."[16] Indeed, although we are still years away from a molecular assembler, the path has been cleared by none other than the U.S. Congress, not a body eager to embrace novelty and change.

A publicity release from the American Institute of Physics, in their *Bulletin of Science Policy News* for July 7, 1999, said this:

One area of research that is beginning to come in for special interest from the White House and Congress is nanotechnology—the study and application of materials, devices, and systems on a scale of nanometers [a billionth of a meter]. At this scale researchers are learning to manipulate individ-

ual atoms, an ability that experts testified could lead to revolutions in materials design, manufacturing, medicine,
electronics, energy, and numerous other fields of human
endeavor. The President's science advisor, Neal Lane, has
rated nanotechnology one of the government's 11 inter-
agency R&D priorities [in planning the 2001 budget].

The United States federal government was then spending about
$230 million a year on nano research, with the National Science
Foundation, Department of Defense, and Department of Energy
the key players, now boosted to half a billion. That's small potatoes by the standards of the decades-long War on Cancer or the
Manhattan Project, but it is an encouraging start. Other nations
such as Japan are also getting into the funding race.

Here is another quote from the Institute of Physics announce-
ment:

> Paul McWhorter of Sandia National Laboratories compared
> the promise of nanotechnology to the first silicon revolution
> in microelectronics, saying this "second silicon revolution"
> had the potential to surpass the impact of the first. "Twen-
> tieth century technologies . . . pale in comparison with what
> will be possible" when scientists can build things one atom
> at a time, said Rice University's Richard Smalley, winner of
> the 1996 Nobel Prize in Chemistry.

Smalley, as it happens, won his 1996 Nobel prize for the invention
of fullerenes—one of the keys to today's laboratory explorations
of a workable nanotechnology. In his talk, Smalley described che-
motherapy he was undergoing as a "blunt tool" killing many
other noncancer cells in his body indiscriminately. Medical nanotechnology, Smalley said, "would allow specially engineered
drugs to target just cancer cells."
So there's nothing foolishly optimistic about this projection of
a new technology able to put things together atom by atom. Fix
your attention on this remarkable implication: a mature version
of such a technology, coupled to powerful computers, to a large

extent replaces workers, management staff, and the factory itself. Given cheap raw chemical feedstocks and a suitable environment, nano assemblers, or matter compilers, will share the ability of viruses to make copies of themselves. When the *factories* are self-replicating, we might stand at the front door of utopia. Unless, of course, they are munitions factories, in which case we might all be doomed. It is arguable that merely human intelligence is insufficient to deal with these crushing prospects. Luckily (perhaps) we can expect a new kind of enhanced intelligence to arrive among us at about the same time as any effective or threatening molecular nanotechnology.

Artificial intelligence

As I mentioned previously, Dr. Hans Moravec, director of the Mobile Robot Laboratory in the Robotics Institute of Carnegie Mellon University, argues that we can expect artificial intelligence to exceed our own by 2040.

In 1988, in his first book, *Mind Children*, the estimated time of arrival had been 2030. Moravec added an extra ten years to take account of a decade's hard-won, hands-on experience—but his new date still attracts vehement criticism from other computer engineers. In *Robot* (1999), he states that "with a firmer launch ramp" he now "projects humanlike competence in [inflation adjusted] thousand-dollar computers in forty years. A slight rise in the estimated difficulty had been partially offset by faster growth in computer power" (p. viii). His reasoning is the kind we have seen already: trends, mapped on logarithmic graph paper. An exponential growth curve makes a straight line up and across log paper, and that's precisely what Moravec found when he did certain dexterous moves with the available data. The number of angels you could fit on the head of a pin, so to speak, had been increasing steadily by a factor of a thousand every twenty years. Given the shockingly tiny size of Very Large Scale Integrated Circuitry, let alone more recent shrinkages, the head of a pin makes a suitable comparison.

Actually, Moravec was describing the quantity of computational power you get for your buck. It worked out to a doubling every two years.

Moore's law again

Moravec's estimate wasn't quite as hearty as that famous speculation known as Moore's law. In 80 years, Moravec pointed out happily, there has been a trillionfold cut (that is, to one 10^{12}th) in the cost of computation. At this rate, if it kept up, a machine equal to a human mind seemed due in supercomputer form by 2010 and on your desk by 2030. And despite the naysayers, the rate of improvement hasn't markedly slowed yet (although, as noted earlier, Moravec now thinks these major advances might take a decade longer). As long ago as 1995, *Scientific American*'s 150th anniversary issue announced that the rate was, if anything, *accelerating:*

> the rate of improvement in microprocessor technology has risen from 35 percent a year only a decade ago to its current high of approximately 55 percent a year, or almost 4 percent each month. Processors are now three times faster than had been predicted in the early 1980's; it is as if our wish was granted, and we now have machines from the year 2000.

In other words, the double-your-computer-for-no-increase-in-price curve had twisted up the chart, and was now doing its doubling act every eighteen months.

Here's a way to extract some sense of the prodigious numbers we're talking about: in May 1996, Tsukuba University in Kyoto, Japan, announced their zippy new superparallel computer, the CP-PACS. It yoked together about a thousand upscale microprocessors to do 300 billion (3 by 10^{11}) separate calculations, or "flops" (FLoating point OPerations), a second. In computerspeak, that's called 300 gigaflops. The estimate rose to 600 gigaflops later that year, with an additional thousand processors added to the array. By December 1996, the United States Department of Energy's Sandia National Laboratory and Intel Corporation, the world's largest computer chip company, announced a parallel supercomputer that had *already* reached 1.06 *trillion* floating point operations (or teraflops) using just 7264 of the planned 9200 Pentium Pro processors.

That is a scary jump in power. We are talking about an array of machines doing a million million calculations each second. When it was finished in 1997, it sustained 1.4 teraflops and peaked at about two teraflops. Not to be outdone, late in 1999, IBM signed a $100 million contract with the American government to build an ultrasupercomputer within five years. Named Blue Gene, a descendent of the Deep Blue machine that unnerved chess grandmaster Gary Kasparov in 1997, it would run a *thousand* trillion operations a second (a petaflop), two million times faster than today's best desktop computers.

It is easy to see that Moore's Law is right on target for the Spike.

Even so, back in August 1996, *New York Times* cyber specialist Ashley Dunn attacked the idea of computing power reaching AI level. True, Dunn admitted, "whether densities double every 12, 18 or 24 months, their proliferation is still remarkable by any standard. The advancement from 2,300 transistors with 10-micron features to the P7's 10 million transistors with 10-micron features is the stuff of technological legend." But he went on:

> What is significant, however, is that the pace of development is slowing. There are barriers now that are so difficult to surmount that advancement from this point on will only come at an enormous price. The end point of the great rush across the frontier is still decades away, but it can be seen.

The crucial problem, Dunn argued, is that to get chip density etchings down under one ten millionth of a meter, cramming in components ever more tightly, calls for such "exotic light sources" as X rays. While there's nothing to stop chip makers using X rays to etch their products, it will cost, and cost a lot.

Two can play at the game of runaway geometric surges, the pessimists as well as the optimists. Both, of course, regard themselves as realists. Dunn predicted exponential increase in equipment costs, almost offsetting the exponential benefits in cost-for-dollar of raw machine power. Another Intel board member, Arthur Rock, had found that fabrication equipment costs double every four years. Intel's new chip plant, devoted to making circuit chips at tolerances of a quarter of a millionth of a meter, cost $1.5 billion. Dunn asked, "Who knows what the next generation will cost?"

Moravec fights back

I passed the objections along to Hans Moravec. What response might he offer his technical critics? He was crisp, standing by the *Scientific American* estimate of a revised Moore's law:

> Richard Wallace taught a computer architecture class at New York University in 1994 and, as a class project, had his students collect data to update the power/cost curve. Rather than flattening, the curve had steepened from doubling every two years, to doubling every 18 months. Can you doubt it, from your own experiences with personal computers in the last decade?

There are still those *technical* reasons adduced by Dunn for thinking that computer improvements might be on their last legs. True, my home PC has gone from a steel-boxed KayPro II in the very early 1980s, with 64K of Random Access Memory and a laughably limited 191K floppy disk for data storage, to my current modest Celeron with 128 megabytes of RAM and eight billion bytes of hard-drive storage. (That's two thousand times as much working memory in less than 20 years. Forty thousand times as much data storage, on a vastly better and quicker medium. A hugely bigger display screen, and in rich color. And, what's more, rather less expensive . . .)

Personal computers, however, are still merely early exhibits from the slow part of the computing curve. Today's cutting-edge components are not just small, they are *very, very small*. Sooner or later, one is bound to run out of profitably accessible space, down there on the etched circuits. Even if there's ample unused room at the nano level, will we be able to push the atoms around quickly enough, and cheaply enough, to continue our dizzying plunge downward into micro- and nanochip utopia, and upward toward the Spike?

Salvation from the labs

Certainly, Hans Moravec assured me. Relax. "The engineers directly involved in making ICs [integrated circuits] tend to be pessimistic about future progress, because they can see all the problems with the current approaches, but are too busy to pay much attention to the far-out alternatives in the research labs. As long as the conventional approaches continue to be improved, the radical alternatives don't have a competitive chance. But, as soon as progress in conventional techniques falters, radical alternatives jump ahead, and start a new cycle of refinement. When short wavelength ultraviolet is no longer good enough, there are X-ray synchrotrons, electron-beams and even scanning tunnelling microscopes. Existing circuits are still so far from the 3-D, quantum electronic possibilities already demonstrated in the lab, and the economic incentives to keep the race going are so huge, that I see no reason to expect things to start to slow down for several decades—and then we have a whole new regime, that will make our current time look like the Stone Age. (And if chip plants cost ten billion dollars, well, that's partly because they have to be so big to make the quantities that the market demands.)"

The latest news from Moravec is even more astonishing. The Moore's law curve returned to its original doubling-every-year by 1997, according to his analysis of computational bang for your buck, and has now swept onward into even swifter acceleration. In 1993 personal computers provided 10 MIPS (million instructions per second), by 1995 it was 30 MIPS, by 1997 it was over 100 MIPS—the brainpower of a housefly. Moravec comments: "Suddenly machines are reading text, recognizing speech, and robots are driving themselves cross-country."

Very well. Stipulate that Moravec's instinct is right, and Dunn's is wrong. (And even Dunn didn't deny that progress might yet continue its wild gallop for decades.) Once we reach the stage of computers as smart as people—whatever *that* means, and it could mean a lot of different things, as we'll see—well, all curves then do roar off the graph. It's the Spike. Moravec was more explicit still in the opening draft passages of his book *Robot,* later omitted: "Exponential growth is very fast, and underlies many frightening scenarios, for instance Malthusian population explosions. The growth we are considering is faster still, because subsequent doublings happen in shorter and shorter intervals. Such a rule can take a mathematical curve to infinity in a finite time—a point called a 'singularity,' which term was first applied to technical progress by Vernor Vinge. Beyond the singularity the curve may have an entirely different character that no simple extrapolation will divine. [. . . T]echnical progress is indeed taking us toward a singularity, probably within fifty years."

Unlike our poor, limited, bone-encased, once-and-for-all, genetically constrained brains, AI systems should be able to continue to double in power, and shrink in price, at an unnerving rate: perhaps every year and a half, perhaps even faster once computers are made by nanotechnology, grown from the molecule up, perhaps even *incredibly faster,* since they'll be designing their replacements and amplifications at truly inhuman speed.

This sort of rhapsodic rhetoric does give sensible people pause. It really has the smack of . . . *military-industrial hype.* It is easy to dismiss, especially when you see how shamelessly its proponents pursue their logic.

Ed Regis finds behind these radical intuitions "the desire for perfect knowledge and total power. The goal was complete omnipotence; the power to remake humanity, earth, the universe at large. If you're tired of the ills of the flesh, then *get rid of* the flesh: we can *do* that now. If the universe isn't good enough for you, then *remake* it, from the ground up."

Spike fever. Is it something we should worry about? University of California, Irvine, physicist and writer Professor Gregory Benford has expressed some doubts: "Most of humanity won't take part . . . If a small segment takes off beyond view, they will still need to protect their physical well-being—necessities, etc.— against the slings and arrows of outraged humanity (and there are always such; envy is eternal). So this juncture will provide the real working surface for change . . . those in the Singularity will be beyond view, anyway."[17] Maybe advanced technology won't take over our lives—but unless we look these options in the eye, while they're still on the Computer-Assisted Design drawing screens, we'll wake up one day and find (yet again) that the shit has already hit the fan. Despite Dr. Benford's assessment that those few who pass through the Spike will be constrained by their dependence upon the rest of us for resources and power supplies, this is not at all self-evident (an objection Vinge himself made in response). The time to anticipate traffic gridlock and backseat teenaged sex was when the first Tin Lizzies bumped down un-made roads. We might just be at that stage in the development of nanotechnology.

Remaking our world

What *can* we do with machines so small you lose them if you breathe too hard? Well, send them inside your arteries to do the job of heart surgeons, for a start. One cynical wit has noted, admittedly, that this is more like a million thudding snowplows banging around inside you than a microsurgeon's lean fingers. Still, in a world approaching a technological Spike, nothing seems out of bounds.

Might we really freeze our heads after unavoidable death, as I suggested in the previous chapter, and later be returned to life? Yes, with the right nanotech. Those billions of dedicated machines will mosey in through your rewarmed bloodstream and cut and splice, repair and replace, make your damaged tissues good as new—better, in fact, since full mastery of the genome will ensure that the reborn are also the rejuvenated. Cancer, recall, is a disorder arising especially in those tissues that constantly repair themselves, so it's the downside of cellular immortality. Fix the repair system and find indefinite longevity.

But why bother? Better, some argue, to have your vulnerable brain scanned and uploaded into a dexterous, robust machine. ("Uploading," by the way, is when data passes from a less complex device to a bigger and faster one. "Downloading" goes the other way. You download information from the Internet to your small desktop computer. The assumption is that minds will one day be copied into machines *even more powerful* than human brains—*up*loaded.)

Scanned at the individual neuron level—of course we don't know how yet, but we *will*, we *will*—you could be backed up, like a computer disk. Die, and you're retrieved and regrown by a version of cloning technology, one in which your stored memories are somehow programmed into the new growing brain. Or perhaps your body and its brain's memories are literally nano-assembled, using the stored backup information and plenty of carbon and other freely available atoms. VR and genome science and nano meet in this vision of paradise (or inferno) on earth, brought to you care of hard science.

While it sounds like grandiose paranoia, its advocates are disarmingly exuberant, cheerfully self-mocking. Keith Henson—a programmer whose youthful sport, with his wife Carolyn, was Recreational Explosives—planned the Far Edge Party. In 200,000 years, his many-copied selves and their friends mean to meet at the other side of the galaxy and swap notes. If the party gets big enough, the vast mass of the bean dip will collapse under its own gravity, forming a black hole: the Bean Dip Catastrophe.[18]

Japes aside, my feeling is that science and technology are feared less for such big unannounced jumps and more for their gradual encroachments on our peace of mind, even if we don't yet consciously understand that this dread is gnawing at us. But dentistry is a technology, guided by science. Would you wish your teeth to rot? The telephone is a technology, and so is mechanical transport. Do you want to walk everywhere, restricted to a circle of twenty or thirty kilometers? Eyeglasses are a technology. Would you prefer to become functionally illiterate in middle age?

Nano around the corner

Eric Drexler's dream of machines petite as molecules has leaped out into the TV news: the other day I watched a news report of a geared gadget powered by a single evaporating water droplet. Not nano, but getting there.

Control over materials at the nanoscale means building machines for our own purposes *from the atom up*. Instead of starting with big lumps of metal and plastic and grinding them down to roughly the right size and shape, nano promises to *grow* tools (indeed, anything you wish) that will exhibit immense precision and strength. Ultimately, and maybe not all that far off, nano control might yield a "black box" that could build you whatever you ask for, as long as you can give it the coded description and a batch of raw chemicals. Computers and nano go together very neatly, especially when the computers are also made of nanoscale machines: cunning calculators comprising invisibly small rods and levers built from carbon composites resembling diamond.

What then? Everyone at the dawn of the twenty-first century is worried about how hard it is for many people to find paid work, even after specialized training, but the sad (or happy) truth is that quite soon humans will not be needed for most of the world's work. There will be *no reason* for most of us to submit to toil, to do anything except enjoy our lives on the ample fruits of the nano/artificial intelligence dividend. How will we spend the endless holidays? Mangoes and ennui, Ed Regis feared in *Nano!* Bosnia everywhere is what worries me. Or teen gangs with nanofactured guns, and literally endless supplies of ammunition and deranging intoxicants.

An even more monstrous calamity is possible. Drexler worried about the Gray Goo Catastrophe, when the nanomachines mutate and start eating everything around them, pumping out sludge (or endless beef steaks). In 1991, a popular book by Chris Peterson (who happens to be Drexler's wife and colleague) and Gayle Pergamit, with Drexler's name added to help boost sales, carried forward the initial Edenic nano vision into a grittier reality. In this book, *Unbounding the Future*, Drexler et al. gave us our first glimpse of what a realistic minting future might look like. Attempting to defuse such fears, he and his colleagues wrote calmingly: "This would be an extraordinary accident indeed," one that could only start if someone inexplicably built "a highly capable device that is almost disastrously dangerous, but held in check by a few safeguards. This would be like wiring your house with dynamite and relying on a safety-catch to protect the trigger" (p. 252). Then again, despite expert warnings, corporations around the world do keep building nuclear power reactors on top of dangerously unstable fault lines . . .

Certainly we will insist that industrial nanomachines must have finicky tastes, depending for their very survival on artful "vitamin supplements" unavailable in the wild—which is to say, beyond the lab or factory. The Prime Directive that Drexler et al. propose is stark: *"Never make a replicator that can use an abundant natural compound as fuel"* (their italics, p. 253). Otherwise we might end up, after all, in the future they mockingly dub *Pollyanna Triumphant:* "Products begin to pour forth. The economy is thrown into turmoil. Military equipment also begins to pour forth, and tensions begin to build. A military research group with more cleverness than sense builds a monster replicator, it eats everything, and we all die" (p. 269).

Oh. Fortunately, the authors add, "This scenario is absurd, at least in part because published warnings already exist." Still, one might be excused for harboring some doubts about the force of this reassurance. We have read no end of warnings about no end of evils since . . . well, since prophets starting issuing warnings from mountaintops. True, discussing the lethal hazards for everyone concerned seems to have forestalled global nuclear war, but deranged nanobuilders will be able to do their dirty work in the garage, alongside the crazed gene engineers who decide to build a bug tailored to wipe out, say, all obese redheads and leave everyone else in perfect health. If so, we'll be spared the excitement of the Spike because, a few years before it's due, "we all die."

No doubt rigid critics will conclude that these diverse predictions of daily life on the rising slope of the Spike are themselves advance examples of the Gray Goo Catastrophe. For my money, they'd be wrong. Whether we like it or not, the Spike is on its way.

But what *will* happen at the Spike?

Vernor Vinge sketched his disturbing theory of the Singularity to a group of interested humans at the February 1995 meeting of the San Diego chapter of the Association of Computing Machinery—a wonderfully nineteenth-century name for a bunch of people hearing, no doubt with some skepticism, that humanity is due to be outstripped by computers. Vinge started with the obvious moves and countermoves. Yes, computing hardware per dollar has been zooming away at an exponential rate, but even exponential swells within specific processes are bound to *saturate:* they don't go on improving forever.

Indeed, they often collapse catastrophically, as animal populations do after a prodigious bout of breeding that outstrips the carrying capacity of the landscape. Vinge hoped that this particular doleful fate was not in store for us. If we avoided it, however, we were in for something marvelously difficult to pin down, because imagination simply fails when change is pushed at a thousand times the customary rate, let alone a million times.

Despite the doubters, technologies with the wind behind them "are usually a superposition of saturating curves of the individual technologies" making up their component parts. As one curve falls away, another takes up the slack. Vinge insisted that current computing trends would continue for another three decades. For example, Internet nodes were increasing by a frantic 30 percent per month. And even that explosive increase could continue for a long time.

Vinge's specialist audience understood the implications of this superposed exponential curve. It wasn't anything as mundane as adding more and more (boring) cable channels to your television reception, or extra cells to the mobile phone system, or even hooking up remote, isolated parts of the world to the global tele-communications network. All of these advances are, in a sense, merely additive. If wretchedly poor people in the heart of some Third World country suddenly gain access to the telephone, to the global positioning satellites, it will improve their lives a little but it won't revolutionize the world in utterly unpredictable ways.

True, the fall of Soviet communism and various other revolts against gray authority were said to have been catalyzed, dynam-ized, by the fax machine, the Internet, even the photocopier. But Vinge was speaking about a grander jump: closing a gulf between animal and mineral, between living brains and silicon or gallium arsenide hardware.

Machines as smart as people

The world telephone network, even with its billions of switches and speed-of-light exchange of data, will never "wake up" when it hits some critical density and find that . . . It's Alive! But arti-ficial intelligence programs run on very swift machines that might do just that. Techno-optimists suspect that human complexity could be mimicked by devices only a hundred or a thousand times better than existing hardware. If this were true, we'd reach human emulation somewhere between 2005 and 2030. (Two thousand five! Twenty oh-five! As Vinge gave the lecture, that was just a decade away.)

Others, sure that the mind and its supporting brain were not quite so easily emulated, concluded that processing is done down at the cellular level of the brain, rather than at the coarser "chunked" level of neural networks. That would give a million times as much power. If so, artificial intelligence able to match the human mind would take longer to arrive on the scene.

And there are some who believe AI just isn't going to happen. The most notable of these was Sir Roger Penrose, Rouse Ball professor of mathematics at Oxford University, theorist of black holes and twistors (don't ask), cowinner with Stephen Hawking of the 1988 Wolf Prize for their joint contribution to our understanding of the universe. Penrose was not an intellectual opponent to dismiss lightly, and he thought that consciousness—and that meant true intelligence, not just its clumsy counterfeit—depended on some as-yet-unknown physics, quantum gravity, which permitted the curious structure of living neurons to do calculations beyond the range of mere computers.

Vinge took this challenge seriously, but made the obvious retort: if there's nothing metaphysical, nothing mystically unknowable in Penrose's purported quantum neurons (and there isn't, Penrose himself asserts), why then, that technical trick can be mastered as well. It's just a matter of doing the research and development work. Maybe the R&D will take extra decades.

The hair-raising point, the really significant consideration, is this: "What do you build five months after that! Or what does *it* build five months after that?" What does the AI do with its new consciousness, its ferocious, self-augmenting intellectual powers?

That is the edge of a technological Singularity, the place when the future starts to go completely opaque. Once a human-level machine takes charge of its own development, with its storage and internal connections and speed doubling every eighteen months or much faster, you get a superhuman-level machine in (historically speaking) the blink of an eye.

When is the Spike due to happen? Some fans of the Singularity, Vinge told his audience, have formed the 2014 Club. "Actually," he added with a smile, "May thirteenth, two thousand fourteen."

And what happens a little after the Spike? Could be a striking event, Vinge supposed. "You could look out to the West and say, 'I don't remember a mountain range out there.' "

Dr. David Brin, sf writer and author of *The Transparent Society*, has offered a comparable comment slightly less apocalyptic but unnerving for all that: "A good parent wants the best for his or her children, and for them to be better. And yet, it can be poign-ant to imagine them—or perhaps their grandchildren—living al-most like gods, with omniscient knowledge and perception, and near immortality."[22]

But remember, one great truth about trends is this: the farther they're projected into the future, the less reliable they are.

Trusting the trends—maybe

The unreliability of trends is due precisely to *relentless, unpre-dictable change,* which makes the future interesting but also ren-ders it opaque. How can you guess what some research physicist or engineer will cook up in her lab ten years down the track, let alone forty? Science, we can agree, is exceptional in its intellectual and very practical shocks, its paradigm shifts, its discontinuities. Extending a current curve into the future is, when all is said and done, nothing but guesswork and faith.

Can we trust these exponential curves at all? They're just ar-tifacts, aren't they, no better than an arbitrary choice of curve fitted to a bunch of data points usually representing quite distinct things. Yes, in a ferocious market economy there's going to be *huge* commercial pressure to make the next processor chip smaller, packed more densely with transistors, a jump or two faster. So the companies that fabricate the gadgets fund research into a swath of possible breakthrough areas. Scads of brilliant, trained minds are now alive and working in their labs. Someone is *bound* to *take the next step,* assuming there is one and they can find it—

Hmm. But that means the data points for computer-power-per-dollar *aren't* really independent of each other. Maybe the curve *isn't* bogus? After all, mightn't rates of change have turned out to fluctuate wildly? What was to prevent a dog's leg from showing up on the graph? How come the exponential curve drops so neatly over the last half century if all the data points are *in-dependent*?

There is a somewhat squelching answer: it's the next simplest curve to try after a straight line.

The simplest line between points on a graph is a linear plot, the straight line that matches equal components to equal intervals (first x, second $x + 1$, third $x + 2$, fourth $x + 3$ ...). So you get, say, 1, 2, 3, 4, 5, and onto infinity, each gap the same as the one before it. The next obvious thing to try out is logarithmic plots, exponentials, which add ever larger components per interval (x^1, then x^2, then x^3 ...). Now you have, say, 3, 9, 27, 81, 243, and so on in a soaring, monstrous surge. A log plot leaves out the xs and maps the powers or exponents to which x is raised: 1, 2, 3, 4 ..., so that last runaway upward curve is transformed into yet another a straight line.

What we're closing in on here is a seat-of-the-pants estimate of the *bogosity factor*. That's an intuitive measure of how much wishful thinking is incorporated in your analysis, blended with balderdash, spin-doctoring, and sometimes a dash of outright chicanery. On a scale of one to 10, the "cold fusion" furor scored very well indeed for bogosity: perhaps 8.9 or 9. (There are still some serious labs, mostly in Japan, that persist in working on it, although nobody in those labs believes they are detecting nuclear fusion—maybe the excess is "zero point energy," or something even more exotic. Skeptics deny there's anything to be seen at all.)

What would win a score of 10? The Flat Earth theory, say. "Creation science," certainly. An attempted revival of the phlogiston model of heat, happily abandoned once chemists learned that heat is not a substance, but just the dynamics of the ceaseless motion of atoms.

The bogosity index is culture-specific, you must understand, and certainly not fixed forever, because our knowledge of the universe is always provisional and open to reframing. Since that's so, since we're not talking *Timeless Truth* (which nobody knows), the great foundational structures of biological evolution, quantum theory, relativity, can safely be assigned a bogosity score of zero. For this year, at any rate. Many scientists rate parapsychology's claims in the high 8s and 9s, and the prospects of nanotechnology in the near future only a little lower. I'm agnostic, at this point. Let's give them both a 5.

Well, what about Moore's law, the doubling of computer power every ... year, two years, eighteen months, whatever ... ? An intriguing index of how relentlessly computing power really has been surging onward and upward is the track record of Big Prime Numbers calculated on available machines. A prime number is one that cannot be divided by any other number except itself and 1, and they have to be sought out laboriously, each one individually. There is no rule to predict when a prime will turn up in the sequence of numbers. In September 1996, Cray Research at Silicon Graphics found one with 378,632 digits. That's a single number which, when written out, would fill about 200 pages of a book like this. Since then, the record for largest known prime has been broken no less than four times, due to the introduction of the Great Internet Mersenne Prime Search. GIMPS is a distributed computing research effort, which can be joined by anyone with a computer. It was estimated in 1997 that a one million digit prime would be found within a decade, but the reality was more astonishing. In January 1998, a 909,525 digit prime was found, close to that predicted million—but by June 1, 1999, a new prime with over *two million* decimal digits was located. This was just a year and a half later, seven years sooner than the original prediction.

Don't be misled into thinking this means the size of the largest known prime, rather boringly, had merely doubled (one million to two million bottles of beer)—the reality is far more shocking. Look at it this way. Isaac Asimov once calculated that we could fill the entire volume of the universe with roughly 10^{125} tiny protons cheek by jowl. Recall that this enormous number is equal to 10 followed by 125 zeroes. Compare our jump from a prime nearly one million digits long to the next known larger prime, which needs more than two million digits. This is a mind-boggling vast jump. It's equivalent, on average, to doubling the size of the original number *every eleven seconds* during the entire sixteen months of the search.

The acceleration trend is not slowing, then, but in fact speeding up. I am indebted for this prime number analysis to Greg Jones, a Silicon Valley aerospace engineer who rejoices in the happy nickname of "Spike." Applying an interesting method he discovered, we can estimate that there's nearly an 80 percent chance that a *three* million digit prime will be discovered by November 2001.

Compare these recent mighty bounds of discovery to the first modest step along the way. In 1951, an early computer found a seventy nine digit prime, which at the time seemed quite an achievement. Since then, a logarithmic plot of new record largest primes shows a remarkably straight line heading for the three million digit prime and beyond. On ordinary graph paper, that history would show an impossibly steep curve: in 1951 the first point is down below one hundred, by 1996 it was up past a third of a million, and by the end of 2001 it might top three million digits. You'd use up a lot of paper to draw that curve.

Moore's law, on this evidence, seems to be holding up impressively well.

Futurology—bogus or not?

Well, then, what about the supposed rising curve of scientific attainment itself toward Singularity, to human-level artificial intelligence, to augmented human brains, to superintellects of fleshy or fabricated matter? That's harder to estimate. Let's also set its bogosity factor, for the moment, at 5.

There are more conservative expert opinions, of course. In June 1996, the U.S. Air Force released what amounts to an updating of that forty-year-old forecast used by Stine decades earlier. In a huge year-long project to estimate a range of future worlds possible by 2025, the study strove mightily to avoid palpable bogosity, while rather preening itself on speculating boldly "outside the box" (its term for the confines of convention) in its effort to capture what it dubbed "The Vigilant Edge."

The Air Force was determined not be caught napping. Under the direction of their chief of staff, General Ronald R. Fogleman, they sought to "generate ideas and concepts on the capabilities the United States will require to possess the dominant air and space force in the future." Over two hundred participants took part directly, comprising fifteen scientists and technologists forming an operations analysis team at the Air Force Institute of Technology, cadets, more than seventy guest speakers, including Alvin Toffler, Kevin Kelly, and Dennis Meadows, "experts on creativity and critical thinking; science fiction writers and movie producers; scientists discussing swarming insects, communication capabilities, advances in energy; experts in propulsion systems; military historians; international relations specialists," while two thousand interested parties were consulted via the Internet.

In 3300 pages of text, the report was organized around six scenarios of quite different 2025s: *Gulliver's Travails,* "rampant nationalism, state and nonstate sponsored terrorism, and fluid coalitions"; *Zaibatsu,* a "cyberpunk" future where "multinational corporations dominate international affairs and loosely cooperate in a syndicate to create a superficially benign world"; *Digital Cacophony,* a kind of pre-Spike United States of high computing power and sophistication, global databases, biotechnology and artificial organs, and virtual reality entertainment, but little order and much anxiety; *King Khan,* a Pacific Rim Sino-colossus dominating a First World sunk in gloom and austerity; the cheekily entitled *Halves and Half Naughts,* where fifteen percent of the world is rich while the rest grieves and seethes with nothing to lose; *Crossroads 2015,* an intermediate epoch after near-term war in Eurasia, with constrained rates of economic and technological growth.

None of these possible tomorrows takes the plunge into truly discontinuous technologies, not even the Digital Cacophony. This is how the planners saw that supposedly extreme "futuristic" case:

> Electronic referenda have created pseudo-democracies, but nations and political allegiances have given way to a scramble for wealth amid explosive economic growth. Rapid proliferation of high technology and weapons of mass destruction provide individual independence but social isola-

tion. The US military must cope with a multitude of high technology threats, particularly in cyberspace. The US world view is global, technological change exponential, and the world power grid dispersed.

It is not a terribly bold vision after all, despite the Air Force's consultation with experts on "nanotechnologies and microelectrical mechanical computer processing advances."

An optimistic prediction

By contrast, a strong case can be made that a Spike is imminent, perhaps within a decade. Daniel G. Clemmensen argues it forcefully. Technical progress is speeding up, he notes, in large part because there are now more trained scientists and technologists adding their efforts together—or perhaps multiplying them!—so that less and less effort is needed to supply the essentials of life, even as the tally of what we regard as essential grows in its turn. Our available instruments thrust our progress forward like sails opened to a wind that has always blown but, until now, has never been adequately harnessed.

Movable type and the printing press are a classic instance of how *new tools* can catapult tens, hundreds, millions of human brains into fresh abilities, untouched cognitive landscapes. Such novel tools don't just disperse stocks of raw data and processed information; they enable human brains to work more efficiently. In recent decades, the computer—first lumbering mainframe, then desktop, now worldwide network—has acted as a similar amplifier or accelerator, and its impact can only grow. These technical innovations are not just patches, stuck onto what ancient cultures always knew. They are *thresholds*, steps into larger habitats, unexplored mental ecologies.

"Between thresholds, the basic driving mechanisms cause the mathematical model to be an exponential with time. Each threshold appears mathematically as an increase in the mantissa." The mantissa is the *fractional* part of an exponent or power to which a factor is raised—2^2 increasing to $2^{2.4}$, say, and then to $2^{2.7}$ until it reaches 2^3. "This is intended as an analogy," Clemmensen adds hastily, "not as a rigorous or even a non-rigorous mathematical model!"

So again we are left asking: if it *is* only an analogy, might progress toward the Spike slow and run out of steam any day now? Clemmensen admits the ad hoc nature of the exponential curve but is adamant, even so, that it means just what it seems to mean:

"The technological singularity will occur at some point in the near future when we cross one of these thresholds. This threshold differs qualitatively from the previous ones, because it will enable us to begin generating the equivalent of more thresholds in very rapid succession. The curve will change from an exponential to a hyper-exponential. This is not actually a mathematical singularity, but the rate of progress will become so fast over such a short amount of time that there is no effective difference."

Hyperexponential! Faster than a speeding bullet, and getting faster all the time, and then some. Machines and minds linked together, a kind of hybrid of current-model human and up-and-running Internet, achieving numerical critical mass (maybe around 2006) that will turn its searchlight gaze back upon its own hypercomplex gestalt and start rewriting itself, correcting its own structural defects, adding to its own capabilities—more memory, more processing power, better programs—and all of this within weeks of ignition. Isn't this the promise of wild change we came in with, the declaration of apocalypse, of the Singularity, of the Spike? Clemmensen makes no bones about it:

"Within a short time, everything that can be known, will be known, and anything that is possible within the laws of physics will be achievable."

It's time to recall Drexler, in 1988, on exuberant expectations: "Nanotechnology will offer fertile ground for the generation of new bogosities. It includes ideas that sound wild, and these will suggest ideas that genuinely are wild. The wild-sounding ideas will attract flaky thinkers, drawn by whatever seems dramatic or unconventional."

So step back. Calm down. Before we have humanity catapulted into the estate of godhood, we might be advised to imagine something rather more limited, if still astonishing: simple home improvement in the twenty-first century.

Redecorating made simple

Around a hundred years before Drexler dreamed of nanomachines, the Scottish writer Robert Louis Stevenson had a darker dream, a tale of captivity and resilience that made him instantly popular: the short novel *Treasure Island*. You might remember the looming, clever, word-spinning figure of peg-legged Long John Silver, and his capture of the dazzled boy Jim Hawkins: "You're a lad, you are, but you're as smart as paint."

A curious turn of phrase, even for "the year of grace 17—." Is anything *stupider* than paint? The stuff just lies there on the wall, mindlessly drying out and cracking and eventually flaking off. But Long John borrowed his metaphor from a different sense of "smart": smartly dressed, bright and fresh in appearance, like a newly painted wooden surface. And in Jim's case (although not, I hope, our own), altogether gullible.

Drexler and his colleagues have reborrowed the expression "smart as paint" and given it an amusingly literal spin. Quite soon after the emergence of routine nanotechnology, they suggest, the watery or oily or plastic emulsions we have known until now as paint will be replaced by a new decorating toy: "smart paint." This is paint that looks back at you, so to speak. Not, luckily, paint with attitude. No, paint that does what you tell it to do, that changes color on command, serves as a wall-sized television display or vibrating sound source, that sucks in light when nobody's around and turns it into electricity, that rolls up and goes back into its can when you tire of it.[23]

And why not? Suppose we replace the pigmented constituents of today's paint with a matrix of molecular machines containing infinitesimally small, stupid computers, mounted on rollers built out of a few atoms apiece. No need to store this stuff as a liquid:

Drexler's "paperpaint" is a woody block, easily cut with a special trowel. Using a nontoxic marker (something like a felt pen, I imagine), run a line around the boundaries of the area you mean to decorate. Put a dollop of paperpaint in the center, smooth it a couple of times with the trowel, and watch the solid stuff spread out smoothly to the marked perimeters, and stop.

At a level a hundred times smaller than the microscopic, a zillion smart paint particles in this miracle application have rolled along, telling their neighbors where they are, until they cover the marked area. If there's a crack or a small hole, they've crawled in and filled the gap. Now they wait for you to tell them what color you'd like, resetting their tunable dyes.

Once it's in place, you can say goodbye to some of the housework, since "sufficiently smart paperpaint," Drexler and his colleagues tell us sunnily, "would shed dirt automatically using microscopic brushes." But is it safe? Might it slide down off the wall in the middle of the night and cover your face, painting you into brightly colored death by asphyxiation? Drexler doesn't mention this rather obvious fear, but one has to trust that adequate and foolproof safeguards will be devised before smart paint is released onto the market.

Where's the energy coming from to run this wall of paint? Do we need an extra socket just for the paintwork? On the contrary. Invisibly tiny machines can capture available light and run off it, as plants do. A more ingenious form of protective surface might eventually cover all existing roofs and streets, converting daylight to a trickle of power from every square centimeter, then pooling that power and storing it at a local level. Here's Drexler's own account: "On a sunny day, an area just a few paces on a side would generate a kilowatt of electrical power . . . [P]resent demands . . . could be met with *no* coal burning, *no* oil imports, *no* nuclear power, *no* hydroelectric dams, and *no* land taken over for solar-power generation plants."

A power supply too disseminated to meter! Of course, as with the murderous bedroom paint, one has to hope that a stroll down the lane could not turn into an electrical firestorm, some kind of arcing and spark-spitting *Weird Science* pavement out of control.

Then again, we routinely walk or drive next to overhead power lines carrying enough scurrying electrons to fry us if the lines fall, yet we don't lose much sleep over that risk (preferring, if we are the worrying kind, to fret about possible long-term hazards of low-frequency nonionizing radiation from those same power lines).

The beginnings of nanotechnology

To build a machine on that scale calls for precise control of matter on the level of individual atoms. We already have that, in (as it were) a small way. Years have passed since IBM's Almaden Research Center in San Jose, California, painstakingly shoved one atom after another into the right place to display the company logo at nanoscale, then went onto draw a globe showing North and South America and portions of Antarctica on a scale of one to ten trillion. Those tricks were done at extraordinarily low temperatures, using machines originally designed to see atoms. Scanning tunneling microscopes (STMs) turned out, rather usefully, to be able to shove atoms around as well as detect their current position. The STM in their name might well be interpreted as Seeing, Touching Manipulators. (Announced by Gerd Binnig, Ernst Ruska, and Heinrich Rohrer in 1982, the STM won them the 1986 Nobel prize in physics.)

Until recently, though, scanning probe microscopes had a major defect: their sluggish speed. Their tips, nosing out atoms with an electric field, were suspended above a prepared surface from a probe looming (as IBM Zurich researcher James Gimzewski rather strikingly put it) like the Eiffel Tower hanging upside down from the clouds above Paris. It worked, but you wouldn't want to try painting the Mona Lisa using equipment like that. This gloomy situation is changing.

In the November 1996 issue of the journal *Science,* Stanford University's Calvin Quate announced a modest but encouraging jump in nano productivity. Professor Quate, together with Binnig and Christoph Gerber of IBM Zurich, invented the atomic force microscope (AFM), which can be used to image molecules inside living cells. His team had built an efficient 16-tip microscope-cum-manipulator, and other researchers were reportedly constructing one with 144 tips. Quate hoped to be "writing 1-cm by 1-mm areas and we'll be doing it very fast." On the nano scale (a ten-millionth of a centimeter) this was an astonishing expectation. More recent work by Quate's team uses nanotubes to even greater effect. In 1998, they announced work showing "the possibility of selective etching and indentation of silicon surfaces by carbon nanotubes mounted on an scanning probe microscope cantilever. The proposed method is very robust and does not require applied voltage between the nanotube tips and the surface. We show that . . . nanotubes are able to extract silicon atoms off the silicon surface."

Back in 1996, the American Institute of Physics's *Bulletin of Physics News* had reported something still no less surprising. An abacus, one of the world's most ancient calculating devices, had been constructed by placing and moving individual buckyballs on a copper substrate, using a scanning tunneling microscope, at room temperature. In effect, they made the world's first atom-scale computer. "The buckyballs (which are big, sturdy molecules)," wrote Phillip F. Schewe and Ben Stein, "act as the counters of a tiny abacus in which low (indeed mono-atomic) terraces in the copper surface constrain the buckyballs to move accurately in a straight line."

Small and perfectly formed, this device is, for obvious reasons, agonizingly slow, constricted by the comparatively vast machinery of its scanner tip. The next step, however, will be to "fabricate arrays of hundreds and even thousands of SAM probes for simultaneously imaging (and repositioning) many atoms and molecules." By comparison, the report noted, single features on a Pentium chip—the kind running most of today's desktop and notepad computers—could swallow hundreds of rows of buckyballs. Once the abacus is refined to accept input and deliver output at the same nanoscales as its components, we will see machines of the size needed for smart paint—and a thousand other far more inventive and important applications.

It goes without saying that tomorrow's computers will not be literally constructed in the form of abacuses. Even so, Drexler and Merkle and their fellow researchers have sketched out in some detail a plausible plan for using "rod logic"—something like an abacus, but using Boolean logic rather than simple addition and subtraction—in place of electronics. Shuttling such minute, rigid rods back and forth, each of them built of just a few atoms and sliding inside channels a molecule wide, a nanocomputer could replicate all the standard features of electronic digital computing.

In a sense, this would be a shockingly retrograde step, like counting on your fingers. Still, due to the Lilliputian dimensions of the shuttling rods, they will fly with blurring speed, and outrun any current machine. Eventually, Drexler assures us, the technical difficulties of controlling electric or optic flows at nanoscale will be overcome, and rod logic will retire, replaced by a molecular version of the machines we know today.

By then, of course, it is likely that new breakthroughs will have transcended these prospects, moving us into the realms of quantum computers that crunch not bits but "qubits," probabilities in quantum space, and far sooner than we had any right to expect. Already, even without quantum computers, the first rampart of human intellectual pride fell when Gary Kasparov—the greatest chess grandmaster of all time—lost by one game in his return match with IBM's Deep Blue computer. For centuries, chess was regarded as a supreme game of genius and strategy, courage and imagination (although the Japanese game Go is even more complex). Now it is no longer possible to doubt that machines can rival the performance of the finest human minds—even if computers are not yet conscious. The world can never be the same again. Technology's rate of change is screaming upward on the graph, heading for the Singularity.

3: Little Things

[C]OMMONPLACE NOTIONS PAINT A FUTURE FULL OF TERRIBLE DILEM-
MAS, AND THE NOTION THAT A TECHNOLOGICAL CHANGE WILL LET US
ESCAPE FROM THEM SMACKS OF THE IDEA THAT SOME TECHNOLOGICAL
FIX CAN SAVE THE INDUSTRIAL SYSTEM. THE PROSPECT, THOUGH, IS
QUITE DIFFERENT. THE INDUSTRIAL SYSTEM WON'T BE FIXED, IT WILL
BE JUNKED AND RECYCLED. THE PROSPECT ISN'T MORE INDUSTRIAL
WEALTH RIPPED FROM THE FLESH OF THE EARTH, BUT GREEN WEALTH
UNFOLDING FROM PROCESSES AS CLEAN AS A GROWING TREE . . .
IN RETROSPECT, THE WHOLESALE REPLACEMENT OF TWENTIETH-
CENTURY TECHNOLOGIES WILL SURELY BE SEEN AS A TECHNOLOGICAL
REVOLUTION, AS A PROCESS ENCOMPASSING A GREAT BREAK-
THROUGH . . . THESE DEVELOPMENTS ARE TAKING SHAPE RIGHT NOW,
AND IT WOULD BE RASH TO ASSUME THAT THEIR CONSEQUENCES WILL
BE MANY YEARS DELAYED.
      —DREXLER, PETERSON, AND PERGAMIT, 1991, PP. 22, 24

Living Green with tailored molecules! Ah, what a wholesome, magical prospect. Shut down the factories, roll in the nano black
boxes. *Technoenvironmentalism.* What an advertising agency's
dream run . . .

It is easy to scoff, and many people will. But Drexler and his
colleagues may well be correct in the long term, even if their
timetable can be questioned as ambitious. Sooner or later, control
of nature at the level of the molecule is likely to provide cheap
and energy-efficient production of food and other primary consumer goods. And that is just the start, which is what makes the
skeptics nervous. Proponents forecast nanomachines that will
replicate without direct human control, scour plaque from our
aging blood vessels, repair damaged brain cells, ultimately replace
biological components altogether.

Start with micromachines, then think smaller

The cover story for the millennial opening issue of *Business Week* was "21 Ideas for the 21st Century," and the fourth was "Molecular Machines Aren't Fantasy. Just Ask the Pentagon." Sidebars took readers to a variety of sources of news on micro- and nanoscale technology of a kind that ten years earlier would have been dismissed—and had been, as we'll see shortly—precisely as absurd fantasies. Here's how the story opened:

> In the 2020s, you may be able to buy a "recipe" for a PC over the net, insert plastic and conductive molecules into your "nanobox," and have it spit out a computer.
> Matter will become software. That's not a misprint ... we'll be able to use the Internet to download not just software but hardware, too. So predicts James C. Ellenbogen, the nanotechnology honcho at Mitre Corp., a Pentagon-funded research center in McLean, VA.[1]

These forecasts emphasize matter-compiled computers, to start with, but the machines these scientists have in mind are the size of salt grains. Their working parts will be microscale, laid down in molecular layers by a process resembling photocopying. Something like that is already available, turning CAD drawings and instructions into full-sized three-dimensional plastic mock-ups of whatever you care to describe in a computer program. But those renderings don't have any working innards, any more than a hologram does (being made purely of light). By contrast, the new machines described by Ellenbogen are themselves essentially circuitry, and that *can* be fabricated by sequential deposition.

Ellenbogen's team has already designed a small interim fabricator a sixth of an inch (five millimeters) long. This is vast in nano terms, but such microelectromechanical systems (MEMS) will help make still smaller machines to make even tinier fabricators. What of Drexler's vision of true nano-machines? Building computers atom by atom, the *Business Week* article concludes, still remains a comparatively distant dream, so Ellenbogen is "betting on molecular electronics for the near term." What is striking about such stories, though, isn't that Drexler's hopes are seen as too difficult to carry through in the next five or ten years—it's that they are taken very seriously indeed, almost with a complacent shrug, for the long term. Miter Corporation's own Web site notes, "Micro-robots made via contemporary techniques for nanomanipulation and nanofabrication are a logical nearer-term step toward the more-difficult, longer term goal of constructing artificial nanometer-scale machinery, such as molecular assemblers."

Assembling molecules

Whatever you want to build, Ralph Merkle has claimed, if you have the plans and the laws of physics don't forbid it, you can make one—using molecular assemblers. With MNT, you can build a skyscraper or a skijet or a tube of lipstick in Blushing Pink or a diamond tunnel to China paved in Penrose tessellations. Every atom will be in the right place, specified by the computer program driving its molecular assembly.

True, you can also expect to manufacture—to *nano*facture—tiny dedicated computers the size of bacteria to control the hues of your smart paint wall, or photovoltaic cells that pave the street and provide your electricity at very little extra cost, or surgical gadgets that swim into your chromosomes and repair those fraying components that help bring on death in all complex organisms. But emphasizing the vastness of the invisible world—all that room at the bottom—soon to be opened by nanotech probes, that was a cardinal error. That was to miss half the point. Nano was a way of remaking the world at our own mesoscale, midway between the atom and the cosmos.

Using diamond, built cheaply carbon atom by carbon atom, we would sit in chairs with a strength-to-weight ratio fifty times better than steel, and if those chairs included smart nano components, rather than just being built by them, why, sitting in diamondoid chairs would be a truly sensuous experience as they molded themselves moment by moment to your best posture.

In nature, diamond can only be made by extreme pressure in the bowels of the earth. With a nano assembler (including the very first one, which would be made out of gigantic proximal probes) you'd be free to pick up a carbon with your tip, hold it in temporary place, get another carbon, sidle up at the correct angle, watch them pop together, go and get another carbon atom . . .

Minting

Let us call this process of nanofacture "minting," from MNT, that clunky bunch of words: molecular nanotechnology.

So you could mint a diamond, legendary icon of ineffable beauty, immemorial permanence, and extravagant costliness, exactly because a diamond is a stable arrangement of one of the commonest elements in the world. Gold and silver, even more ancient symbols of wealth, might retain their price, because they cannot be constructed from more plentiful atoms. True, nano scavengers might float eventually in the oceans and select such atoms one by one, popping them into a bag for later collection at the shore. Such are the prospects of material profusion in a world permeated by small, smart machines.

But for the immediate future—twenty, forty years—basic nanoscale assemblers were on the drawing boards, Merkle assured us, awaiting a lot of detailed engineering, but arguably ready for the major development push that would yield the new minting technology. "As good as a license to mint money," clear-eyed cynics say today of peppercorn-level electronic media permits issued by governments to radio and television proprietors. Nanotech promised to make good that metaphor. Guided by their ancillary computer programs (we'll come back to these), its assemblers were the mint itself. If you had a nanofacturing system and a supply of raw materials, you could mint . . . well, Merkle was here to tell you: *anything not forbidden by the laws of physics.*

Bypassing biology

Nano's proponents have thought this project through in depth. A critic might object: wouldn't it be preferable to mimic (or conscript) existing biological machines—cellular ribosomes, etc.—to do the difficult work of growing your products from the molecule up? After all, that's what living systems do right now. Why start over?

Because, Drexler pointed out in *Engines of Creation* (I paraphrase loosely), evolution did its design work by accident, blindly, wandering in design space and constructing each new story atop any previous attempt, however pitiful, that didn't collapse immediately. This is not the best way to maximize one's end products. Using molecular assemblers designed instead by the human mind, he noted, engineers "will build entire computers smaller than a synapse and a millionfold faster."[2]

They would be clunky things, for all that, laughably nineteenth century. The only good thing to say about them is that their moving parts would be made directly out of atoms, and atoms don't wear out, or feel friction, and because they are so incredibly small they can be shifted back and forth almost as readily as a light beam. And, unlike most conventional lathes and conveyor belts, they could easily make copies of themselves.

Drexler's assembler

This was Eric Drexler's original conception of a self-replicating assembler.

Suppose you needed a gadget with ten thousand moving parts. You could design one in which each unit part was made of just a hundred atoms, for a total of a million atoms. The machine might fit into a box two hundred atoms long, fifty wide, a hundred deep—still at nanoscale.

What do those mechanical parts do? Some form a shell to shield the assembler from the outside world, which would probably be a fairly selected environment to start with. Others form conveyor belts—literally! *Conveyor belts!*—to haul in items of feedstock: carbon atoms, hydrogen atoms, chlorine atoms . . . Still others make up motors, and instruction tapes, and rod-logic computers to interpret and implement those tapes. At the heart of the thing, still others comprise a stubby robot arm, with small motors able to swing it back and forth and up and down, like an infinitesimal scanning tunneling microscope (STM) tip. And near its tip, yet more gadgets will change the tools held poised above the work surface.

Since this robot arm is so much shorter and lighter than the meaty one hanging from your own shoulder, it can whip back and forth with astonishing precision and agility, easily a million times a second: picking up a selected atom from the conveyor belt (as instructed by its tiny onboard computer), carrying it to the next place in the fabrication sequence, positioning it with excruciating precision, dropping it in place with a pulse of electric charge, returning to the bin or belt for a new atom.

You can get this machine to make a copy of itself simply by providing it with a set of taped instructions recapitulating the set of movements needed to build it in the first place. The new assembler, exactly the same size as its parent, would emerge from its side, sealed against the environment until the two separated, the short end of the child emerging from the wide end of the parent.

For efficient replication, Drexler notes, it might be best to have dedicated units containing several assembler arms and several jigs to hold the work pieces firmly in place. All up, perhaps 150 million atoms. To be cautious, make that a billion. At a plausible pace, such a system will copy itself in a little over fifteen minutes, just as a bacterium does.

Assembler factories

Nobody wants to build a new car or a skyscraper one atom at a time, however fast the individual atoms are handled. Which is the great virtue of self-replication in your assemblers. To start the workday, you need to brew up a pond of nano-rich soup. Starting from a single assembler and plenty of raw materials, by the end of ten hours of exponential growth, you'd have a stock of some 70 billion minting gadgets. Reprogram them to build something else the same size, and you'll have another 70 billion nano-objects floating in the pool within a quarter hour. And so it goes.

Why a pool? Because it makes sense to build in three dimensions. Weaving flat sheets with assemblers handling a million atoms per arm, you can turn out a hundred layers a second—but even at this dizzying rate you'll need a year to fabricate a slab a meter thick. Do it in depth, growing outward from a "seed" surrounded by a volume of porous scaffolding, bathed by a fluid of feedstock and perhaps prefabricated components such as carbon fibers, and you can build very much faster. So: prefab subunits at leisure, and stockpile. "Molecular assemblers will team up with larger assemblers to build big things quickly."

Drexler's presentation was breathtaking, but apparently feasible for all that. To build a large rocket engine (designed by computer on a modularized plan), you flood a large vat with "thick, milky fluid"—an image not lost on psychoanalytically inclined critics—rich with prepared assemblers programmed in advance "by making them copy and spread a new instruction tape (a bit like infecting bacteria with a virus)." At the center of a base plate is a seed containing "stored engine plans, and its surface sports patches to which assemblers stick."

This is one of the most crucial features of nano design, not unlike the developmental apparatus of living embryos growing from a primal fertilized cell. As the first assemblers dock with the seed, they learn how they are oriented to it and each other. Procedures are downloaded into their simpler brains. "A sort of assembler-crystal grows from the chaos of the liquid. Since each assembler knows its location in the plan, it snags more assemblers only where more are needed."

Unlike an embryo, however, we see something like a skeleton or nervous system organizing itself first, and only later filling in the connective tissue and organs. Even that isn't exact, because finally this skeletal scaffolding will withdraw from the growing lattice, its nano units broken down and flushed away, perhaps to be reprogrammed and recycled for the next task.

Before that happens, the construction proper takes place. Buzzing assembler arms lock carbon atoms in place, arranged into the strong configuration known as diamond fibers or perhaps fullerenes. "Where resistance to heat and corrosion is essential (as on many surfaces)," Drexler notes, "they build similar structures of aluminum oxide, in its sapphire form."

It's illuminating, though, once you notice it, how many gestures at programming and instruction you find in Drexler's pages. Consider the "seed," a nice device but chockful of knowledge that needed to be put there by prodigious human effort. Well, not all of it—perhaps some of what it ordains to the mindless assemblers is data gathered by "disassemblers." This data needs to be shrunk by compression routines under intelligent command. Maybe this task can be turned over to a no-brain number cruncher, as the compression of two-dimensional graphic files already are. But maybe it can't.

Unbounding future capitalism

In 1991, in *Unbounding the Future,* Drexler, Peterson, and Pergamit offered a modest scenario: no pixie-dust nano-Genies, no tabletop "anything boxes." Rather, an early twenty-first-century Mom-and-Pop company, Desert Rose Industries, "a diversified wholesale manufacturer of . . . furniture, computers, toys, and recreation equipment." Far from the magic nano-Genies implied by skeptical mockers (and some hopeful enthusiasts), this is "a plumbers' nightmare of piping," ponds and containers labeled "CARBON FEEDSTOCK, PREPARED PLATINUM, SIZE-4 STRUCTURAL FIBERS, and PRE-FAB MOTORS." As well as the visible building ponds, there's an underground warehouse stocked with many hundreds of tons of prepared building blocks, made elsewhere and shipped in by magnetic levitation subways running at aircraft speeds.

Mixed together, these components form a syrup ready to be

organized by "block assemblers." Once activated, the key gadgets lock together with their neighbors, divide jobs between them, "use sticky grippers to pull specific kinds of building blocks out of the liquid," and "use their arms to plug them together," all under the master control—and this strikes me as a very important detail, sometimes overlooked—of the plant's desktop computer. It's all clean, powered by solar cells coating the roadway, exhaling not noxious pollutants but "an updraft of clean, warm air."

Extraordinary prospect? Yes. But not nearly as difficult to swallow as the medium-term and long-term promise of a mature minting technology, the kind that might place an *haute cuisine* foodmaker in your kitchen or a free-to-the-public matter compiler on every street corner. Yet those, in turn, might be commonplace in the decade or two after nanofacture is exhibited but before we get all the way to the Spike.

Half magic

As ever, this is the hard thing to take in, to comprehend: that at the Spike itself, which might be sooner than a century away, might indeed be only *half a century* away, or even sooner, things will be so different that we, here and now, have trouble even trying to conceive of them. Really, when all is said and done, that single assertion is the very heart of what this book is trying to convey. We can't map the path of the Spike, can't analyze it with the customary tools of history and reason, because by definition it will create a horizon to prediction. It will make the future, in a strict sense, unimaginable. (Which doesn't mean we won't try to peek around the edges.)

Writer John Barnes, in a cautious and mathematically informed paper entitled "How to Build a Future," reminds us that radical technology always contains an effectively "magical" quality.[3] Not literally, of course, nothing supernatural or sorcerous, but from the point of view of people in the previous generation it genuinely seems magical, incomprehensible. This is the meaning of Arthur C. Clarke's celebrated Third Law of prediction: *Any*

sufficiently advanced technology is indistinguishable from magic.[4]
Barnes suggests that:

Each new surge is 90 percent what you might have expected
from the last one, plus 10 percent magic (in its Clarke's Law
sense). So from the viewpoint of 1920, 90 percent of the
gadgets of the (roughly) Manhattan Project through Apollo
Project boom would be imaginable (indeed, some, like TV,
were abortively available in the previous boom). But 10 percent (lasers, nuclear power, transistors) would be absolutely
incomprehensible—magic. [By Surge Three, early in the
2100s], 52 percent of significant new technology in the culture we're imagining must be stuff we currently would not
find comprehensible.

Realistically, the world should be half magic.

While knowing this perfectly well, Drexler has learned nonetheless to moderate his public claims, to keep (as he recommended
to his supporters at the Seattle "Nanocon" in 1989) "the level of
cultishness and bullshit down," repressing the talk "about wild
consequences," however technically defensible those might be.

Much of it, though, was damage-control management strategy,
a deliberate decision not to frighten the horses. Drexler, according
to Carl Feynman, has become "a much more careful fellow in
presenting his ideas over the years. Initially—he was twenty-one
or so when I met him—he had these crazy ideas, and stood too
close, and waved his arms, and talked a little bit too loud . . . I
guess he's found that amazing ideas come over better if you're
extremely calm."

The proposed engineering, all the hard work by many collab-
orative specialists and generalists, came together in the formida-
bly dense pages of *Nanosystems: Molecular Machinery,
Manufacturing, and Computation,* Drexler's 1992 *magnum opus.*
Some experts, on the other hand, remain skeptical. Dr. Julius
Rebek, an MIT expert in the chemistry of self-assembly, was
scornful. "This is not science," he was quoted as saying dismis-
sively, "it's show business." Merkle found that objection muddy

and hard to grasp. Does Rebek deny the feasibility of positional control of single atoms? Surely not, since IBM Zurich had already achieved just that, and at room temperature. Of course, current scanning probe systems, leviathans by nano standards, are a far cry from the nifty assemblers upon which minting will depend.

And perhaps this brings us to the heart of the dispute: Merkle and his allies argue in favor of hard, detailed preparatory work, theoretical as much as experimental, that will break the ground for later advances: just the kind of detail that thickens the 556 pages of Drexler's *Nanosystems*.

In particular, computational chemistry—a blend of abstruse quantum theory and computerized analysis and imaging—sidesteps or shortcuts many traditional maneuvers in chemistry. Give a computational chemist "a molecular structure that's supposed to do something useful," Merkle notes with a certain breeziness, "and they'll happily model it. Sometimes they come back and tell you that it works and it's a great design. Sometimes they torture it in horrible ways and it behaves in some strange and unexpected manner . . . at which point, it's back to the drawing board." He adds: "The design of an assembler will require extensive analysis on a computer if it is to be developed in a reasonable time frame."

Still, even if minting became a manufacturing reality, would it be the inexpensive boon promised by *Nanosystems* and similar exploratory books? After all, building tiny specialized modules at the molecular scale and sticking them together might not be terribly cheap. Besides, perhaps it's a waste of time and effort, as nature already does it for us. Why go to the trouble of constructing a nanobuilt chair, when you can get a tree to grow you a nice slab of wood?

Merkle's predictable reply: today we use a huge, costly thing— a chip fabrication facility—to build a tiny, fairly cheap thing—a computer chip. Spread the costs of making that individual chip among its brethren, and one sees that silicon computer chips are not all that cheap, costing up to millions of dollars per kilo. "In sharp contrast," he says, "we have examples of very small things (cells) that make other very small things (other cells) and do so inexpensively (perhaps a dollar a pound). Nanotechnology," he adds, "proposes to use self-replication to achieve low cost." And that, indeed, is the key to minting's promise.

Assemblers assembling assemblers

If the assemblers are able to make copies of *themselves,* the unit cost drops through the floor. And anything else they can be instructed to produce is also likely to be inexpensive, not much more costly (once the programming is amortized) than the raw feedstocks of elements that they process into useful macroscale objects.

As estimated in *Nanosystems,* in 1992 values, the "chief anticipated feedstock materials" are the abundant elements carbon, nitrogen, oxygen, and hydrogen, all of them surrounding us in the atmosphere but currently expensive to snatch from that source. Bought in bulk industrial quantities, they are priced around 10 cents per kilogram. Other elements needed frequently—silicon, phosphorus, sulfur, fluorine, and chlorine, all from the second and third rows of the periodic table—are "reasonably inexpensive." Catalytic metals such as platinum do cost an arm and a leg, but since a kilo can be used to process ten billion kilos of material a year, its unit price is vanishingly small.

What about energy? All of these guesstimates have an eerie circular quality, because once efficient minting comes onstream so will cheap photocells and other inexpensive sources of power. Even aside from this likely state of affairs, Drexler notes that nanofacture will be a net energy *producer,* so that power captured as a by-product might *exceed* the costs of the feedstocks.

Wastes? Neither heat nor water given off by nano processes are expected to be significant. Not much land will be required for the typical nanofactory, because "the exemplary manufacturing system produces its own mass in product in less than an hour," suggesting very much diminished ratios of land needed per unit of production. This is Drexler's considered conclusion:

[T]he basic cost of production, here taken to exclude the
costs of development and distribution (and of such impon-
derables as taxation, licensing agreements, and insurance),
will be almost wholly determined by the cost of materials.
The relevant materials presently cost ~0.1 to 0.5 $/kg; this
range can thus be taken as an upper bound on the basic
cost of producing a mature molecular manufacturing technology base . . . low enough . . . to be competitive in making
a wide range of products. [*Nanosystems,* pp. 433–34]

Well, this is modest enough. Drexler has indeed taken his own
advice: no bullshit, and forget for the moment that gleaming
diamondoid one-piece spaceship motor rising out of the swarming engines of creation in the nano pond. Still, that lovely thing
remains the goal of the nano planners. But only the long-term
goal. The path to reaching it might involve several distinct but
converging tracks.

Robert J. Bradbury draws a useful distinction between three
distinct orders of nano minting: self-*replication,* self-*assembly,* and
precise *molecular assembly.* Bacteria and higher organisms already
do these tricks, which "soft/wet" biosynthesis methods will emulate. The harder step will be moving to "hard/dry" assembly,
which will allow us to build mansions and spaceships out of diamond and sapphire. The economic and cultural impacts of these
three varieties are not identical. With programmable self-
replication, Bradbury notes, costs of most parts drop to the same
low level typical of timber or food products—lower, in fact, since
they can be compiled from waste. With programmable self-
assembly, most expensive human labor costs are eliminated
(along, of course, with jobs and wages). This is no more startling
than Apple's "automated" Mac plants or Japanese robotic car-
assembly factories, except that now the modular parts have built-
in "intelligence," knowing how to assemble themselves. Just
mastering those first two steps will pay off handsomely.

So the really hard aspect of minting, which requires diamondoid mechanosynthesis chemistry, is only needed to build excep-
tionally strong, light structures. It isn't necessary to compile food

or the mechanisms for energy production from the sun, wind, waves, or for most housing. Building tall structures, Bradbury observes, is limited by the strength of materials like bone, hydroxyapatite (teeth), seashells, and bamboo—somewhat lower than steel and concrete.[5] But semi-intelligent pieces of steel or concrete could self-assemble, or bamboo-beamed houses could "grow" themselves.[6] Interestingly, methods for achieving these small interim breakthroughs were worked out several decades ago, before nanoengineering had been suggested.

Santa machines

I first got wind of the general idea of self-replicating assembly in 1978, three years before Eric Drexler's first published paper announcing nanotechnology. In those days it was called "Santa Claus Machines," and depended on macroscale automatic factories (*not* nano) conceived as independent explorers of the solar system by the man who named them, Theodore Taylor, a former Princeton engineering professor. His imagery was startlingly similar to the nano vision, but on a larger scale:

> It's possible to imagine a machine that could scoop up material—rocks from the Moon or rocks from asteroids—process them inside and produce just about any product: washing machines or teacups or automobiles or starships . . . Once such a machine exists it could gather sunlight and materials that it's sitting on, and produce on call whatever product anybody wants to name, as long as someone knows how to make it and those instructions can be given to the machine. I think the name Santa Claus Machine for such a device is appropriate.

They're always building spaceships, these wonderful machines. The link is more obvious in this case, it's true, since Taylor expected his Santas to fly through space and infest lifeless rocks, turning them into consumables for the benefit of their owner-builders back on Earth (or maybe in space colonies).

Indeed, the possibility wasn't lost on NASA, often regarded by the public, not always unjustifiably, as a nest of conservative reptiles lazing in the Florida sun beside rusting launch pads from the glory days of lunar exploration already a generation gone. Actually, in 1980, NASA sponsored a Summer Study by the Replicating Systems Concepts Team, which yielded a 1982 report, *Advanced Automation for Space Missions* (NASA CP-2255), edited by Robert A. Freitas Jr., who in 1999 would publish the first of three detailed volumes on nanomedicine. It was an ambitious scheme, long since moved to the back burner. Previous studies had proposed autoreplicating lunar devices with a mere 90 to 96 percent "closure" or self-sufficiency. The summer team's theoretical design goal, Freitas observed, was a self-replicating lunar factory grown from a hundred-ton seed landed by rocket, able to get *all* its energy from the environment, and to build 100 percent of its offspring's parts. (Information might still have to be radioed in from Earth, for fine-tuning.)

> This *specifically* included on-site chip manufacturing . . . Of the original 100-ton seed, we estimated the chipmaking facility would mass 7 tons and would draw about 20 kW of power . . . One hundred percent materials closure was achieved "by eliminating the need for many . . . exotic elements in [earlier designs, resulting in] the minimum requirements for qualitative materials closure . . . This list includes reagents necessary for the production of microelectronic circuitry."

That was distinctly *not* nanofacture, but still it was capable of building its own chips! Some space technology experts believe that if NASA had pursued this research program, we might by now have robot-constructed stations all over the Moon and Mars, and maybe on some of the asteroids, exploring, sending back vast amounts of exotic data, perhaps mining and fabricating stuff to be fired back home by solar-powered electromagnetic catapults. Perhaps this ambition will be fulfilled, much more cheaply, using neat, grain-sized machines swarming with nanobots, fifty years later.

Several years before that NASA Summer Study, I had read about Taylor's original scheme in *Spaceships of the Mind,* a book based on a BBC2 television series written and presented by the superb science journalist Nigel Calder. Interestingly, Drexler's magisterial *Nanosystems* fails to mention his predecessor's name. Perhaps this is not so surprising. It is as if Alexander Bell, urging his colleagues to consider the possibility of a, what could you call it, a *telephone,* had been anticipated by some ingenious fellow suggesting that we could communicate from one side of town to the other by bellowing through mighty megaphones mounted on old lighthouses.[19]

Controlling the nanos

The enchanting possibility of Santa Claus Machines took fire in my mind, and I imagined great fleets of these tireless computerized workers trawling the ocean floors for mineral nodes, turning out endless supplies of cheap goodies for the lazing utopians we would all become in this regime of abundance. Until something went wrong with some Santa's programming, of course, and it began eating all the whales, or turning out an unstoppable flood of toasters . . .

In the miniaturized Drexler version, this would be the dreaded gray goo catastrophe, and the defense would be to maintain your assemblers in a state of extreme stupidity, utterly controllable from the outside, their onboard molecular computers savagely restricted in what they might be permitted to organize. On top of that precaution, Drexler urged the development of a variety of "active shields," a kind of nano antibody system, shoes shined, scout hats on, always prepared. Others have suggested widespread "police nanos," a kind of *blue goo* or nanotechnological immune system. Eugene Leitl, a cryobiological chemist, quickly discerned "an intrinsic asymmetry between the Blue and the Gray, which makes Blue always a losing party. Gray is just mindless brute-force autoreplication, with zero checking (light weight) and lots of evolutionary activity (adaptation, which Blue cannot afford, or else it will occasionally change color, and nucleate very militant Gray). Blue needs to know where Gray is (lots of cruising, sophisticated sensorics → added complexity/bulk), and identify it as Gray: docking (time, dedicated docking device → bulk) or other communications shebang, cryptographic authentication (time, circuitry). Blue needs means of termination, which require (a) a highly energetic event (energy storage requirement), (b) violation of the inner vacuum chamber integrity (disassembling device to puncture the hull). If Blue can terminate Gray, so can Gray Blue. Termination of Blue is a side effect of Gray's standard metabolism. Both must autoreplicate."

This thoughtful, rather dashing critique attracted a swift response from Anders Sandberg, a Swedish neuroscientist. Multicellular life presents a different story, he noted. No single-celled selfish replicator has arisen to eat all other life. (That would be a kind of truly feral bacterium; a virus can't copy itself.) Must gray goo replicate faster than immune-system blue? Not if the environment is swamped with simple blue variants, alert for undesirables. Can gray goo adapt faster? "Blue goo could be trained to recognize allowed systems and attack everything else." How does blue tell enemy gray from friendly blue? Well, blue is smarter, more complex than gray, with cryptographic surface structures that might cost extra to make but pay for themselves in security. Finally, is it easier to kill bad goo than to make more? Not at all. Blue can gum up gray, wall it in; gray needs to find quite specific chemical feedstocks in its environment, assemble copies.[20]

But as Leitl mentioned, so also does blue police shield goo. We have a kind of spiraling ecological contest. What's more, the true loser in such a struggle is life itself—and not only the human life that blue goo is trying to protect from all manner of gray and other hued nanoenemies. Intuitively, he notes, while the opponents will eventually reach a standoff equilibrium, "it is just too costly (additional load) to protect bubbles of ecology with heavy-duty perimeters of Blue, and a high concentration of Blue within your body is not exactly fun either. With inert or actively protected suits a human should be able to move among Gray goo for a limited time, but you can't exactly clothe a forest or a cornfield that way."[21] The bleak solution might be a forced transition to some form of *post*biological existence—a topic we'll explore in chapters 4 and 5.

Training your goo

This question of control and intelligence, of the level of precision needed in instructing and copying and securing replicating assemblers, is an issue that has never ceased to vex me. The debate between skeptics and optimistic nanoists (to use the term as a nonderogatory shorthand) seems to depend on what is sometimes called "tacit knowledge."

Tacit, unrecorded knowledge (or its absence) is the bugbear of most current attempts to develop artificially intelligent computer systems. Cultural knowledge is very largely composed of silly little items that nobody has ever written down in a book, plus all the unspoken rules for connecting them together. We gain access to this knowledge by stumbling around in the world making guesses and mistakes, starting as infants and continuing until the day we die.

At a level deeper than learned culture, we are rich in tacit know-how about space and time and causality, tricks—effective ways for dealing with the world—"hardwired" into our ancestors through blundering, blind evolutionary processes. To this extent one can regard today's rich diversity of life as a mélange of algorithms written in the genetic code and expressed in huge active bundles of protein, the outcome of a three-billion-year planet-sized computer processing run.

So how will these little nanites be programmed to do their job of building a starship in the backyard pool? Where do they obtain their tacit knowledge? It's no good pointing to biological gadgets such as cellular mitochondria and ribosomes and crying "existence proof!," unless you can specify or emulate all the subtle algorithms they encode.

Well, you can start by assuming a large number of programmers, aided by a Very Big Smart Early Twenty-first Century Artificial Intelligence that isn't yet quite autonomous, to number-crunch the specs for your starship. Drexler has done the numbers for us, and they are enormous.

To make a block with sides just one-thousandth of a millimeter long (a micron) takes about a hundred billion (10^{11}) assembly operations. Putting such cubic-micron blocks together to compile a kilogram object takes a further million billion (10^{15}) block assembly operations. How do you design something like that? A thousand designers each making one decision per second would eat up a century in planning just a single cubic-micron. Obviously, most of nano design has to be done without direct human attention. That calls for very good computing facilities, and excellent self-controlling programs.

So much for the design part. Then ask: how is this template conveyed to the trillions of nano fabricators doing the backstroke in the milky pool of goop that is a minting nanofactory? Radio? Infrared? Batons (or quipu, as the Incas would have done) handed physically back and forth? Recall that the computer running these gadgets is programmed by "bumpy polymer molecules . . . like a punched paper tape."

How are these independently operated nanites brought into global harmony? *Unbounding the Future,* you'll recall, depended on Desert Rose Industries' desktop computer to instruct the matrix of block assemblers. Is there a precise, preprogrammed cascade of fabricators (a shockingly complex, data-intensive proposal)? Or do the assemblers follow waves of chemical gradients, a standard organic living creature make-a-meat-body method that is far less precise? Not in a goop pool, one might suppose. But one might be wrong.

Again, recall the "seed" and the unfolding ranks of fabrication steps it initiates. Drexler proposes a tree of such assembly steps: "If a product is to be made from micron-scale blocks, this amounts to specifying the transport of the blocks of the correct types to the correct assembly workspaces in the correct order, and specifying the sequence of motions to be executed by manipulators in putting the blocks together." Luckily, many nanomechanical structures will be fairly simple, based on highly regular, repeated structures.

What's more, some of those micron-scale units will be computers as complex as today's supercomputers. The growing edges of a fabrication can contain a vast number of these programmed bean counters, perhaps electronically linked by thin metal cables an atom or two wide. The computer modules can eat these wires when they're not needed any longer, as some spiders digest their web threads at nightfall and spin it anew the next day. Another method of keeping assemblers in touch, sketched in some detail in *Nanosystems,* uses acoustic signals, binary code pulsed to actuators by bursts of pressure in the solution where they float. As a bonus, these waves of pressure provide both control *and* power. The mints, we might say, pump up by pumping up the volume.

So perhaps simple minted objects can be characterized by available CAD-designer programs. As Drexler points out, each micro-sized module, or higher-order component, does not require a separate subroutine to describe it and its placement. Rather, it is akin to the digital sentence created when a computer program *calls* a subroutine to perform some repetitive operation. "Writing a program with 10^{15} lines of code would be out of the question today on grounds of complexity," Drexler admits. A thousand million million lines of code would daunt even Bill Gates. On the other hand, though, "running a program that executes 10^{15} instructions is less extraordinary, amounting to a month of supercomputer time."

Modules make it easy

Minting, in short, is based on *chunking:* the process of making truly stupefying numbers of small identical components—readily described in a CAD computer program—and joining them together in ascending ranks of complicated modules.

Drexler and his collaborators sketched just such a prospect: "If a car were assembled from normal-sized robots from a thousand pieces, each piece having been assembled by smaller robots from a thousand smaller pieces . . . then only ten levels of assembly process would separate cars from molecules."

Perhaps this passage ought not be taken literally as a sketch of how to build up from atomic scale. Rather, its ten levels might be seen as a way for the nonspecialist reader to grasp two distinctly nonintuitive factors: first, just how atrociously many atoms there *are* in a car (roughly, 1,000,000,000,000,000,000,000,000, 000,000), and second, how easily this number can be scavenged together when you're using self-replicating assemblers that grow at an exponential rate. I'm not sure that this is all there is to it, though, since Drexler adds: "Molecular assemblers build blocks that go to block assemblers [which] build computers, which go to system assemblers, which build systems, which—at least one path from molecules to large products seems clear enough."

This is the Lego toy model of nanomanufacture. It implies that sarin gas, automobiles, and raw (or cooked) steaks can be compiled in a layered hierarchy, gluing one atom (or molecule, or large chunk) onto another, by following a detailed template.

Compare this with living replicators, the single existence proof available. Genetic algorithms in planetary numbers lurched about on the surface of the earth and under the sea, and indeed as we now know deep within it, for billions of years, replicating and mutating and being winnowed via the success of their expressions—that is, the bodies they manufactured, competing for survival in the macro world. At last, the entire living ecology of the planet has accumulated, and represents a colossal quantity of compressed, schematic information.

But living creatures do *not* entirely *embody* that information. The genome's rather sketchy information—a meager hundred thousand genes—gets unpacked and amplified by borrowing the benefits of two rich external sources.

One is the information-dense environment itself. A linear string of amino acids, the one-for-one expression of the gene sequences that encode it, folds up spontaneously into a protein whose efficacy depends precisely on the new "emergent" folded surface. Parts are tucked away and hidden, others brought together into fresh configurations. The patches left exposed on the tangled surface are what do the chemical job. Such spontaneous self-assembly is partly due to the quantum laws governing atoms, and partly a result of what we might grandly call the spacetime surround of the protein, the constraining fact of living in three spatial dimensions and one of time, curved in a way described by Einstein's tensor equations.

Second, there's been a history of "Darwinian" selection processes at higher and higher hierarchical ("chunked" or "holonic") levels. This kind of selection history, it's now thought, explains how an absurd abundance of neurons at birth gets winnowed in infancy as our brains encounter the limited shapes of local reality. Trial synaptic connections are discarded in droves until we end up just a shadow of our former selves—but a leaner, meaner, more dedicated version, adapted by explicit experience to our parochial surroundings.

Spontaneous order

Depending on self-assembly and Darwinian shaping, when you think about it, isn't terribly surprising. The cake is always far more than the recipe. Remember, a one-cell human embryo uses a mere 100,000 genes to kick-start a self-assembly that ends up with a living creature made out of quadrillions of dedicated cells. How can life manage that awesome trick, without specific instructions guiding the process at each step?

It does so by using compression tricks. Modularity is one—those quadrillions comprise variants of just 254 distinct types of human cell. Then there's segmentation, in which subassemblies are repeated (like the human backbone). They are guided by gradients of chemicals, like fish nosing through the currents for food, instead of by precise individual instructions. And so forth. Indeed, complexity theorists argue that there's more to life than

such data-compression tricks. They suspect that spontaneous or-
dering principles also restrict, and encourage, the range of per-
missible outcomes. If so, we don't yet have a clue (as far as I
know) what the nano equivalents of those rules would be.

Nano artifactors, too, will use the environment's elaborate, unpredictable informational richness, the kind that goes into unpacking a genome and building a living creature. Life has already
explored and exploited it in evolution's truly vast "drunkard's
walk." Trillions upon trillions of variants have been generated,
and then culled out by the pressures of brute survival. While
evolution has been brainless and criminally wasteful, a world of
complexity has been created and winnowed by those vast numbers of variants walking through design space.

After all, as mentioned, the brain also wires itself in such a
fashion, *without needing precise wiring diagrams in the DNA recipe*.
It takes twenty years to build and program a natural human-level
intelligence, even though all the elements are being assembled by
bionanofabricators as fast as they can manage it. And don't forget
that all the cultural information that has modulated our innate
grammar templates is stored in extended knowledge systems of
immense complexity—other people's brains, libraries, work-
places, all our artifacts, the reshaped environment itself—far sur-
passing anything the Human Genome Project holds stored in its
completed files of one typical human's DNA sequence.

Is it likely that nanosystems, designed by human minds, will
bypass all this Darwinian wandering, and leap straight to design
success? John K. Clark, a Miami electrical engineer and nano
enthusiast, has explained why that might be feasible. He argues
that with minting, "If you have a good description of an object,
then you could make another one, and if you don't, then nanotechnology can examine the object and get a detailed description."

Rather than "writing" the huge number of coding steps needed
to specify a complex object, we might scan a desired object at
the atomic level and record the three-dimensional coordinates of
each atom or molecule. Then a zillion teeny nano assemblers will
allocate and time-share the job of making the smallest compo-

nents, then joining those into next-biggest chunks, and up the
ten or so steps to a gleaming, atomically precise Consumer Thing.

Will simple nano work?

This is an amazing suggestion. Just record every atom's position,
and then copy what you've scanned.

How many atoms was that again?

How much memory does this call for? How fast can it be
accessed?

Well, the process will be going on simultaneously in many
parallel locations and levels. We must assume that data is stored
in a zillion independent—but intercommunicating—nanites. We
can also assume that the scanners are using excellent compression
tricks, which saves a lot of storage. After all, many items of instruction take this form: "Find a carbon atom; link it to the previous carbon atom on the left; find another carbon atom; repeat
five trillion times."

In *Nanosystems*, recall, Drexler proposes molecular "tape" storage. This molecular string would use "polymer chains with side
groups of two distinct kinds," where "a partially fluorinated poly-
ethylene molecule can store two bits per carbon atom by representing a 1 with an F atom and a O with an H atom." With this
nifty tape and its reader, you can store a handy 10^{21} bits per cubic
centimeter, and access your data "substantially" faster "than those
of macroscale disk storage systems." Moreover, it's plausible that
by the time we need such mind-numbing amounts of mass storage, new holographic techniques already in the pipeline might
add an extra wrinkle or two. John Clark has suggested that a
powerful, very brief pulse of laser light might pass through an
opaque object. Using the foremost fringes of that pulse of light,
and a reference beam, it might be feasible to record as a hologram
the positions of every component *without* evaporating the
original.

Still, it's not clear that we aren't begging major questions by
bravely positing "a good description of an object" to be matter-

compiled inside a mint as complex as a factory. Consider Drexler's blithe estimate:

> Small molecular manufacturing systems are comparable in complexity to modern automated factories, both in parts count and in organization. Since macroscopic automated factories fall within the range of present human design capabilities, their nanomechanical equivalents should as well.

But is any minting system worth its salt really no more complex than a robot car assembly factory? For that matter, folded away inside its organizational edifice, doesn't such an automated factory contain quite a lot of macroscale decision arbiters (also known as "people"), working with an abundance of tacit knowledge at the same time scale as all humans and robots to date? Nanosystems will function very much faster, so they will have to operate, in Kevin Kelly's telling phrase, "out of control." Unless our nano fabricators are guided by artificial minds, or computer-augmented human ones, they will need to be programmed by people with all-too-human limitations.

Darwin to the rescue

I'm inclined to think we'll get interesting results faster through "Darwinian" simulations inside computers. That approach would use what's called "artificial life" ("A-life"), based on cellular automata (CAs), and genetic algorithms (GAs). For our purposes, the details are unimportant, but it is impressive that a whole artificial ecology, designed by Tom Ray, already lives inside a computer, breeding and mutating in just the ways Darwin would have expected. It's called Tierra, and it has provoked an enormous amount of interest among computer scientists, AI experts, and even biologists.[7]

Let's call this kind of thing *stochastic (or random) emergent order*. Consider what happened when Tom Ray first switched on his Tierra artificial life simulation and boggled at what he found coming out of its digital ecology:

> I started with a creature 80 bytes large . . . because that's the best I could come up with . . . I let the program run overnight and the next morning there was a creature—not a parasite, but a fully self-replicating creature—that was only 22 bytes! I was completely baffled how a creature could manage to self-replicate in only 22 instructions without stealing instructions from others . . . To share this novelty, I distributed its basic algorithm onto the Net. A computer science student at MIT saw my explanation, but somehow didn't get [i.e., receive] the code of the 22 creature. He tried to recreate it by hand, but the best he could do was to get it to 31 instructions. He was quite distressed when he found out that I came up with 22 instructions in my sleep!"[8]

The point is this: a swarm of A-life competitors in a simulation landscape might evolve solutions to nanotechnology design problems far faster than any carefully thought-out exploration.

In May 1996 Tom Ray—at the Osaka Advanced Technology Research Center—and his colleagues in the Tierra Working Group released a variety of rudimentary artificial "animals" into a worldwide ecology created inside a hundred linked computers, accessed through the Internet. Tierra's code is now available for others to play with and breed their own critters. Some fears have even been raised that such artificial code-creatures might escape and corrupt the rest of the Internet. Techno-optimists even suggest that such a breakout might be the beginning of a genuine autonomous AI, born and bred on the Internet and running in the background, stealing machine operating time unused by their owners and programmers . . .

Darwinism is not restricted to actual flesh-and-blood living creatures, then. It isn't even restricted to the kinds of computers we already possess. There are other possibilities on the horizon, too. Recall the quantum computers mentioned at the end of chapter 1. Using quantum computations, many "virtual" calculations would be carried on simultaneously. It is plausible that a quantum machine might be able to search through "Darwinian design space" for the best possible nano designs, and produce the answers very swiftly indeed. This apparently absurd idea is getting closer to implementation as the first experimental quantum gates are being built. I suspect, though, that we will have rudimentary minting even before we get quantum computers.

Computing with DNA

Another way to tap emergent order—to get design work done fast and "for free"—is to use artfully coded DNA strands. Such DNA computations borrow biological chain reactions to create truly staggering numbers of variant DNA strings within a few hours. When they are mixed, certain strands stick together, and the clumps can be winnowed to find the optimal coded solution.

This process is not the wild fantasy it seems. It's already been used by its inventor, Leonard M. Adelman of the University of Southern California, to solve a difficult problem, the "traveling salesman" search for the optimum route among a number of cities. Princeton's Richard J. Lipton went further, showing that DNA computing can solve otherwise intractable problems. "Because a test tube can hold on the order of 2^{60} strands, you have available a huge number of parallel computers, more than we could ever dream about in a silicon world."[9] In January 2000, Lloyd Smith, a University of Wisconsin-Madison chemistry professor, announced in *Nature* a simple computer with 100 trillion synthetic DNA strands secured to a solid surface rather than sloshing in a tube. The problem solved had only 16 possible answers, but the computer did reach its solution. Meanwhile, a Princeton University biology professor, Laura F. Landweber, has built an RNA computer that searched 512 possible solutions.

Plainly, these experiments are just the starting point. Impossibly complex algorithms generated by such means could control biochemical nano fabrication. If so, the fine-grain detail will very quickly escape our understanding and scrutiny. As a result, we will be obliged to depend for reliable results (and our safety) upon trial and error. Nano minting might need to be done only within impeccable confinement protocols, perhaps overseen by enhanced AI or AI-augmented human watchdogs. That's the downside of the astonishing capabilities promised by programmable minting systems. What's to stop some demented or rash brat from reprogramming his nano-Genie to emit city-engulfing clouds of lethal sarin gas, or to pick uranium atoms out of the sea and stockpile it for later fun times?

If we accept Drexler's own argument, we'll need to maintain strong regulatory authorities. They would license Genies capable of producing dangerous outputs only if these are restricted to sealed assembler labs or factories. This defeats the utopian dream of a Genie in every pocket, but I find it hard to see any other way to forestall planetary doom. Certainly none of this suggests Civilization Made Simple by Scan-and-Follow-the-Dots Nano Minting.

Nanotechnology on the Net

I discussed some of these issues on the Internet with people holding a special interest in nanotechnology, such as John K. Clark. He had tried to pin down just what it is about minting a car that is so different from today's conventional manufacturing. An automobile factory depends, as we've seen, upon the explicit and tacit knowledge of many intelligent people with complex skills. How would minting differ? Clark offered six answers. I wasn't immediately convinced.

The parts a car factory uses are very expensive, he observed, *while the parts that nanotechnology uses (individual, commonplace atoms) are very cheap.* True enough, but this already assumes that nanofacture can use atoms effectively without a sublimely complex assembly protocol, based on extremely ornate design programming, or by brute-force scanning of a prototype. The former, arguably, requires human programmers. The latter is unproved and purely speculative.

A car factory uses many thousands or millions of different types of parts—lathes, welders, jigs, screws, bolts, sheets of curved glass, on and on the list extends. Teams of workers must learn how to operate all of them. By contrast, nano uses at most 92 different parts (the naturally occurring elements). Indeed, almost everything we know is made of fewer than twenty elements, and living creatures use fewer than ten. But is this correct? By the time we're at level three or four of Drexler's Ten Steps from the Atom scenario, presumably we are back to thousands of different specialized sub-units. Some of them, in fact, are screws and brackets, although on an incredibly small scale.

The multitude of components a current factory uses are fragile but we are unable to damage the fundamental units nanotechnology deals with, which are atoms. Well . . . big solid pots are easy to mint, steaks might be immensely more tricky to program. Simplicity might be appropriate when you're minting something like a rocket engine, and perhaps a car, but a scanned and reassembled steak is *part of a cow.* Its fundamental modules are not really atoms but copies of fleshy cells, of many different kinds. (Maybe we can ignore minor differences, because the taste buds probably can't tell. Still, I'm not sure you can build up the connective tissue of a steak layer by layer, or even through and around a foamy "skeleton," without the whole thing going badly wrong somehow. "Waiter, there's a fly in my steak.")

None of the parts a factory uses are absolutely identical, but atoms have no scratches on them to tell them apart. By the time we're well into chunking up the ten steps, "identical" parts will have started to diverge again, especially if they're put together by anything resembling chemo-gradient-guided self-assembly. But I'll readily give ground on this one, since the chunks are, after all, coded by a rigid algorithm.

Nanotechnology can manipulate matter without ever leaving the digital domain. You deal with rods 27 atoms long, or 28 atoms long, but never 27.5601334 atoms long. Again, if we can specify adequately at the digital level, well and good. But how many bytes was that to build a car, specified at the atom-by-atom level? Granted, the crisp digital nature of minting is certainly a point in favor of building from the lowest level up . . . if we can.

Finally, most of the parts a factory uses are very complex. Nanotechnology, by contrast, is like building with Lego blocks. Structures can be built level upon level to arbitrary complexity. It's easy to develop an algorithm to examine any Lego object and then build a duplicate, although it's much harder to find an algorithm that will do the same with a car. Of course one can readily imagine copying a molecular *Lego* design, but it is the monster algorithm for connecting rank upon layer of Legos into a car that's exactly what is at stake. The architecture has to be obtained either by prodigious coding *ab initio*, or by some kind of scanning technology—fabulously advanced magnetic resonance imaging, perhaps, or destructive laser burns that evaporate a steak or a Porsche, layer by layer, while noting where each atom used to be (a sort of *Star Trek* transporter-beam notion).

Or maybe you lower a car into the goop pool and allow a trillion assemblers to pull it apart atom by atom, sending back a record (suitably reduced by those clever compression algorithms) to a filing system in some humungous computer. Drexler suggested this approach in *Engines of Creation*: "A nanomachine able to do this, while recording what it removes layer by layer, is a *disassembler* . . . [A] nanocomputer system will be able to direct the disassembly of an object, record its structure, and then direct the assembly of perfect copies."

On the other hand, to take optimum advantage of nano fabrication requires that we *won't* be using a scanning protocol, since a faithfully atom-by-atom scanned Thunderbird will have all the drawbacks—sexy though they might be—of its original. And if we wish to build a new, improved ergonomic Green car from the atom up, *we're back to needing intelligent, costly programming.*

I don't see how this can be avoided, except by waving the Magic SuperIntelligent Handy Obedient AI wand. That, of course, is hardly out of the question, as we accelerate into the Spike. Still, expecting an intelligent computer to do the dirty grunt work for us might turn out to be akin to the wistful dream bunny rabbits depend upon: that TransRabbits without floppy ears will provide all their food for them. And that's true, since we TransRabbits with our big, strenuously cultivated farms do just that (inadvertently)—until one day we turn up to inject some of the bunnies with unfriendly, lethal calicivirus, as scientists have done recently in Australia, in the hope of obliterating most of the feral rabbit population in a very short space of time.

Manufacturing for free

But do we really need an intelligent machine to run our mints for us? Perhaps this is too pessimistic. Daniel G. Clemmensen, a computer systems architect, has observed: "We already do top-down design, from the level of city planning down to the level of individual circuits within a microprocessor. Why is this qualitatively harder when designing the smallest components from atoms?" What's more (and perhaps this is the key to the best, most transformative aspect of minting): "Even if every single component must be designed by a human, we still get 'free' manufacturing. The design effort is a very small part of the cost of a consumer item."

Minting, of course, should be exactly like copying a CD-ROM or a computer program. The first CD of the latest best-selling record album might cost $10 million to make. The second, and all the other 100 million, costs pennies per unit. Ditto for Windows 2010 and all the computer games that spawn themselves in legitimate and pirated copies throughout the civilized world in a matter of days.

The earliest simple desktop minting machines will be unlike computers and MP3 and DVD players, however, and more akin to the recorded media those machines implement. Nano fabrication systems will be able (when loaded with the right program) to copy *themselves*.

As Clemmensen points out, the impact of design costs effectively shrinks to nearly nothing with nanotech, because those costs are amortized over every unit that grows a salable copy of itself thereafter. Emphasis on *salable*. For a time, therefore, we can expect that patented mints will contain elaborate self-crippling or throttling routines to *prevent* self-replication. Proprietary houses will sell algorithms for furs (no animals died to make this coat!), diamonds as big as the Ritz (boring, Freddy, my house is made of that stuff), steaming chicken soup as only a mother can make it, high-octane gas for the T-bird you just minted, and supplies of sarin gas strictly for allied nations.

Eventually, though, as sure as taxes—which, ironically, might no longer be part of the grand social scheme—canny hackers will break the locks and unleash a blizzard of pirated mints. Within months, wild nano will be loose in the street, psychotics will loose mints that overnight fill the streets with thousands of tons of fried eggs and copies of Mona Lisa with a mustache, panicking governments will—What? Fight back with airborne computer viruses designed to shut down the operating systems of these rogue Salt Mills? Really, the mind boggles.

A transhumanist philosopher at the London School of Economics, Nick Bostrum, has looked at the possible nonconvergence of superintelligent machine and mature molecular nanotechnology:

> If there is nanotechnology but not superintelligence then I would expect technological progress to be very fast by today's standards. Yet, I don't think there would be a singularity, since the design problems involved in creating complex machines or medical instruments would be huge. It would presumably take several years before most things were made by nanotechnology.
>
> If there is superintelligence but not nanotechnology, there might still be a singularity. Some of the most radical possibilities would presumably then not happen—for example, the Earth would not be transformed into a giant computer in a few days—but important features of the world could nonetheless be radically different in a very short time, especially if we think that subjective experiences and intelli-

gent processing are themselves among these important features.[10]

And one or other outcome might be so by . . . 2015. If Merkle and Drexler are correct, 2040 tops. To imagine a world galloping up the slope of the Spike, regard Drexler's quiet, madly sane anticipations in *Engines of Creation*.

Nano and AI would go hand in hand, he thought. A machine that might emulate a human brain could be minted out of compact circuits, wires thinner than the axons used by our brains, operating more than ten million times faster. It might be "an assembler-built block of sapphire the size of a coffee cup" shackled to a tangle of optic fibers and cooling pipes. And these machine minds will not be social isolates; they'll work in fast teams, running real-world experiments as fast as we can now simulate such things. The result: automated engineering complexes forcing technology "with stunning speed." Swiftly, "many areas of technology will advance to the limits set by natural law."

We move forward to stand on the edge of a cliff. Back away with your eyes averted if you must. Here is the medium-term future as Drexler's analysis portrays it: a world with replicating assemblers, able to make anything their programs instruct them to fabricate. Artificial intelligence systems will gain technical prowess and be able to understand human languages, and, even more important, our wishes. Placed in command of energy, feedstocks, and assemblers, such an AI, Drexler admits, "might aptly be called a 'genie machine.' What you ask for, it will produce."

Genie machines. Really, it doesn't matter how long the breakthroughs take to come. Even if Drexler and Merkle and Clark and Clemmensen and the rest are hopelessly optimistic about the timetable, there seems to be absolutely nothing in current science or technology that says it can't be done. And when it is done, we'll start ripping up along one of the Harry Stine power curves, accelerating into a future that's literally beyond today's imagination because its complex weaving of the known and the as-yet-unknowable evades the best calculations we can make.

Nano will take us, will *fling* us, into the Spike.

4: Minting the Spike

HENCEFORTH, IT IS HUMANS AND HUMAN FLESH THAT LAST OUT THE YEARS, NOT THE MECHANICAL INFRASTRUCTURE. OUR BODIES OUT-LAST OUR MACHINES, AND OUR BODIES OUTLAST OUR BELIEFS. PEOPLE WILL OUTLIVE THIS "REVOLUTION"—IF SPARED AN APOCALYPSE, HU MAN INDIVIDUALS WILL OUTLIVE EVERY "TECHNOLOGY" THAT WE ARE CAPABLE OF DEPLOYING. WAVES OF TECHNO-CHANGE WILL COME FASTER AND FASTER, AND WITH LESS AND LESS PERMANENT CONSE-QUENCE . . . THE CONSEQUENCES OF GENUINE INTELLECTUAL FREE-DOM ARE LITERALLY AND RIGHTFULLY UNIMAGINABLE. BUT THE UNIMAGINABLE IS THE RIGHT THING TO DO. THE UNIMAGINABLE IS FAR BETTER THAN PERFECTION, BECAUSE PERFECTION CAN NEVER BE ACHIEVED, AND IT WOULD KILL US IF IT WERE. WHEREAS THE "UNI-MAGINABLE" IS, AT ITS ROOT, MERELY A HEALTHY MEASURE OF OUR OWN LIMITATIONS.
—BRUCE STERLING, VIRIDIAN MANIFESTO, 3 JANUARY 2000[1]

You won't find a conclave of Nobel prize winners in physics and chemistry gathered at Princeton's Institute for Advanced Studies to discuss the onrushing Spike. Yes, and it's hard to imagine a special session of the United Nations called to deal with the un-settling prospect of the Singularity just forty or fifty years down the track, if not sooner. These folks will certainly get together to chew the fat about human contributions to global warming and ozone depletion, so it's not as if they lack an interest in the medium- to long-term fate of the planet, and its currently dom-inant species. No, it's simpler and sadder than that: the idea of the Spike hasn't penetrated yet.

Even more than its several components—nanotechnology, ma-chine intelligence, precise genetic engineering, control of senes-cence, all the rest of the package—the Spike is a topic left to enthusiasts and special interest groups, none of them, naturally enough, certified specialists in the barely emerging field of Singularity Studies.

So you tend, instead, to find the most lively discussions, mixed with waffle and wistful dreaming and shockingly bad poetry and gung-ho computer-nerd free market libertarianism, on sites freely available to anyone with an Internet connection, an inquiring mind, and a forgiving heart: the fairly rigorous newsgroup *sci.nanotech,* for example, moderated by JoSH, who in real life is Dr. J. Storrs Hall of Rutgers University,[2] and open-slather subscription e-mail lists such as those run by transhumanists and Extropians, groups of the like-minded that tend all too easily, and unfairly, to attract the appellation "cultish."[3] If the tenets of the Extropians (Perpetual Progress, Self-transformation, Intelligent Technology, and so on) strike you as ... utopian, a touch unworldly, Ayn Randish, even *cloying,* it is worth remembering that nobody else seems to be paying attention to what surely will be the largest and most momentous transition in all human history—the shift, indeed, from *human* history to *transhuman* and then *posthuman* time scales.[4]

In a rather postmodern mode, everything you find among these voyagers is text and image, overlapping layers of conversation and meditation and pointers to the salient Web site *du jour,* coming to you on your screen in the plainest of vanilla e-mail formats. The language of these ghostly exchanges can range from the dense, obscure but friendly—

> Now (sloppily written) the neighborhood is $(+-0^*E+V,$ $+-1^*E+V, +-2^*E+V, +-3^*E+V, \ldots, +-S^*E+V$ or simply $(+-1^*E+V)$, i=0 ... S. Hmm, let's check: the first term is V, which is the 1d neighborhood ... Seems okay.

—to the learned—

> We cannot expand faster than light, and inside a bounded volume (like spacelike slices of our future lightcone) the Bekenstein Bound seems to hold. This implies that expo-

nential growth of *anything* only is possible in the "short" term (which could be *very* long), then it has to grow as t^3 or slower (assuming a Minkowski spacetime).

—to the convenient (if you're into this kind of thing)—

*Event/Announce: NANOSTRUCTURES: Physics & Technology In: St. Petersburg, Russia From: Mon Jun 19, 2000 To: Fri Jun 23, 2000 Deadline: Tue Feb 01 Tue, 2000 Online: http://www.ioffe.rssi.ru/NANO2000/

One can spend hours a day for months at a time roaming the Net reading this kind of thing, or at least downloading it, flipping swiftly across the surface, saving some to the hard drive for later savoring, dropping a brief comment back to the list if a topic snares your attention, dumping the rest without remorse into the wastebasket before staggering away from the machine, limbs locked and creaky, eyes burning. Luckily, in what follows, I've saved you the effort.

An intense debate on the Extropian e-mail list considered arguments for and against the plausibility of Drexlerian nanotechnology, and whether it will lead to a Singularity. That debate highlighted most of the crucial issues one meets head-on when confronting the prospect of the Spike. Let's look at a digest of the controversy (one half of which has now been revised, enhanced, and placed on the World Wide Web as "Nanotechnology without Genies").[5]

The topic was opened by a man I'll call "Augustine," after the pessimistic Bishop of Hippo, a fourth-century saint of the early Christian church convinced that human possibilities are shackled by sin. Today's Augustine is a crisply articulate engineer who was the second person to join up as a senior associate of the Foresight Institute but increasingly questions the utopian ease of Drexler's projections.[6] His chief opponent I'll call "Pelagius," after St. Augustine's great rival. Pelagius believed in the fundamental goodness of human nature; his reincarnation on the Internet is a buoyantly optimistic libertarian capitalist.[7]

Engines of skepticism

"I was one of the first to buy into Eric [Drexler]'s vision," Augustine announced, "and I have spent much of my time in the last ten years thinking about how to make it happen. But the more I think about the logic of it, the more convinced I am that transhuman intelligence will not look like people expect it to look, and will not have the effects people expect it to have."

Augustine was not alone in admitting to such doubts about the Spike. He found support from social scientist Dr. Robin Hanson, who was spending a lot of his time investigating the "Great Filter" hypothesis. This was a rather grim analysis of a curious, apparently remote astronomical fact of life that might end up telling us a great deal about the likelihood of a successful transition through the coming Spike.

Hanson argues as follows: despite considerable effort, we have not yet found the smallest trace of evidence for intelligent life elsewhere in the universe. Yet there are many reasons for expecting life, even intelligent life, to be plentiful throughout the galaxies, and detectable by us. It follows that any previous alien cultures which have reached our stage of technological proficiency might well have tipped into self-destruction. The Singularity could be on the technological agenda, but its fruit (if this mathematical analysis is valid) might end by poisoning us all.

Countering these hard-edged subversions of their goals and aspirations—Transcendence, in a word—were Pelagius, nano designer Forrest Bishop, Dan Clemmensen, and others. Augustine was not impressed by what he took to be their wishful thinking. He insisted, in his clearly phrased, slightly imperious and therefore irritating fashion, that the feasibility of molecular nanotechnology must be considered from within the context of real-world economies. We need to think first about today's ways of making and distributing goods, then about tomorrow's transitional methods as minting becomes part of the production chain, and, only then, about the days and decades *after* tomorrow, when (or, rather, if) Genie machines emerge from among the smoke and mirrors.

Why should minting be free?

Can we really expect routine near-term minting to give us free goods for the asking? Why should that be so?

After all, carrots and beefsteaks are already produced by the manipulation and assembly of single molecules—performed by the machinery of cells and organs—but they don't arrive free at our doors. They must be grown, harvested, transported, marketed. So if factory owners replace lathes and cutting torches and X-ray lithography beams by machines as small as nature will permit, why should this fact alter the point-of-sale values of goods?

Augustine did not assert, implausibly, that nano would make no difference. He did insist that we think through the precise nature of those differences, and the exact requirements which must be met before Drexler's utopia can come true.

Be irreverent, he urged. "When Ralph Merkle says diamondoid material will be as cheap as potatoes, ask: why potatoes? Why not silk, ivory, mahogany wood, orchids, or caviar?" Might it not prove to be very expensive to make consumables one atom at a time in a replicating system? This challenge was quite shocking. Still, the point was intriguing. We already live in a partially automated economy, and it's becoming more cybernated every week. Robots in Japan build cars very nearly from the raw materials up, but we don't see dealers giving them away, or even selling them for ten percent of the old hand-built price. If Nissan minted their cars in goop pools, would they suddenly decide to make them a gift to anyone who asked for one?

Augustine counseled what he called "calibration": matching one's speculations against known analogies. Is autonomous machine intelligence really likely by 2020, as some claim? Consider the delays and flaws and bugs and limitations of products as ballyhooed as Windows 95 and Seattle's competitors. "How can anyone say that the human mind will be ready to shrink-wrap in that time frame?"

Finally, he undermined the whole airy idea of Genie machines, Santa Claus machines: "If a genie machine is defined as an entity that can make anything—whatever it is told to make—does a genie machine already exist? I would say yes: the economy as a whole is a genie machine. Now, could anything smaller than the entire economy make *anything*?"

True, assembler arms can whip back and forth fifty million times faster than those of a woman on a process line. Yes, infinitesimal rod-logic brains could hold algorithms able to direct those small robots through a series of programmed routines. But to build an entire car by minting it means, in a sense, replicating and condensing not only the entire automobile plant (as Drexler had acknowledged, and as I stressed a moment ago), but also the mines feeding it with bulk materials, and the miners or perhaps the highly skilled operators of its automated mining machines, and the generators pumping in power, and their technical staff, and the trucks hauling raw stuff in and hauling processed stuff out, and the truckers driving them, and the sales yards to get them to the customers—

Growing diamonds on trees

Whoa. Pelagius wasn't accepting *any* of this critique. Why will minting be as cheap as potatoes rather than silk? But silk is only expensive because it's assembled by worms that need exquisite handling. Besides, the creatures aren't very good at making the stuff, having been cobbled together by blind Darwinian processes. Evolution could hardly be said to have optimized the worm to produce silk.

No, neo-Darwinian mutation and selection ensured a far more broadly defined task: getting silk worm genes into the next generation. As for Merkle's potatoes, this excellent foodstuff is a byword for cheap consumables due to an accident of biological history. It's not that potatoes are simpler or stupider than the silk worm's spinneret glands, but that the reproductive strategy they stumbled on, helped by hundreds of years of artificial selection at the hands of farmers, is to make *lots and lots* of new tubers, while silk moths only need a *tiny* amount of silk to get through their self-replication cycle.

Diamonds could indeed grow on trees if any living system had chanced upon the process of clumping carbon atoms into the right tetrahedral geometry to build a diamondoid, and if that trick had turned out relevant to preserving their genes. (Oysters, after all, build pearls, which are not despised, even though it's just a damage-control side effect for the mollusk.) So the significant difference is that minting will be a *designed* process, replacing blind mutation and struggles across peaks and valleys on fitness landscapes by insightful human ingenuity.

Attend to historical and economic context? "None of our previous technological advances improved us, they improved our tools, but in the Singularity we *become* our tools and that is fundamentally new." Pelagius makes an interesting jump here, the kind that Augustine had been at pains to block. One moment we are discussing assembler arms in molecular nano fabricators, the next we are already inside the black hole of the Singularity, coextensive with our artifacts, our minds uploaded into silicon substrates, the flesh left behind with the rotting potatoes and lazing silk worms. There's exhilaration in the air, it's hard to rein in the horses of imagination and bold conjecture.

Paying for the diamond trees

Okay. Augustine is not going to pour on that particular parade. He can agree that some kind of momentous, unprecedented event is imminent. But, he notes, calling it a "singularity" implies that we cannot understand it, which indeed is Vinge's postulate. Why set out from that limiting assumption, that foregone grim conclusion?

Besides, there's nothing new about internalizing our tools. This is what human culture *does*, in myriad ways. As for those diamond trees, well, they'd only grow in certain environments, and still have a price—just as oranges do. Just as everything does, just as you can't get rid of friction in physics.

Assume we have molecular manufacturing, in the sense that we can grow (some) products out of diamondoid materials.

Nanites assemble them according to our specifications, like termites building a hill according to the specifications that are wired into them. These nanites are going to require programming and design, just like any other fine-grained apparatus. Some programmers and designers will do their jobs supremely well, and produce masterpieces of whatever they are making; other programmers and designers will give less attention to their tasks, and they will produce cheaper, generic diamondoid products.

The economy will continue to adjust itself so that typical working people can afford typical mass-market products, and wealthier people can afford carriage-trade products.

Molecular manufacturing will emerge within the same capitalist economy that we live in now. Factories will still be factories. They will require elaborate buildings and millions of dollars' worth of specialized equipment. They will employ biotechnologists with rare and expensive skills. Factories will still be owned by investors who want to get their investment back. They will produce products for the market, and buy inputs from the market.

This argument is one key, it emerges, to the consequent debate. Minting, for Augustine, will be just one more adjustment within the economy, not a disastrous dislocation.

Won't the market be abolished, once nanotechnology is achieved? Hardly. Programmable assemblers ("nanites") could only comprise a Genie machine, Augustine insists, if they replicate the logic and process structures not merely of a factory but of an entire macro-scale economy, a whole world in a bottle, replete with workers and managers and immense amounts of expert knowledge. Even then, he says, they would not be able to build you an object such as, say, a full-scale ship, because the magnitudes involved are absurdly disproportionate—as if we were to ask the current Japanese economy to build us a complete new planet.

How do you pay an artificial intelligence?

It's not the feasibility of nanotechnology Augustine doubts, but its alleged freedom from the laws of the marketplace. Here is his second key argument:

> Let's define a Genie as an entity with at-least-human intelligence and sensorimotor ability, who works for free. (This is distinguished from a genie machine, which is defined as an entity that can make whatever it is told to make . . .)
>
> Nanotechnologists assume that Genies will exist. That's what distinguishes Drexlerian Nanotechnology from ordinary technology.

So it's not actually *size* that counts, apparently, and it's not all that *assembling things atom by atom*. No, what Augustine can't swallow is the necessary prelude to minting, "the idea that entities with at-least-human intelligence will do our bidding." Why *should* they, after all? If you and the entire factory were miniaturized one morning by an annoying Mad Industrialist, would you knuckle down dutifully and do the bastard's bidding? Hell, no. "There are no Genies and never will be." Perhaps machine intelligence, AI, will develop, and even surpass human minds. That's not the point, which is that no uncoerced entity as intelligent as a person will work for free.

The details come later

Let's step aside once more and delve back into *Nanosystems* and its painfully dull outline of the shape of things to come. How do we get There from Here, to the world of assemblers in the goop pool from the astonishing, truly hair-raising but nonetheless prosaic Scanning Tunneling Probes and Atomic Force Microscopes in the IBM labs and similar sites around the globe? It's one thing to prod a single heavy atom back and forth on a flat substrate, even to jostle atomic beads in the tour de force of a molecular abacus, and quite another to build *useful things* with an assembler made out of a million atoms. Or, indeed, to make that assembler in the first place. How *do* we do it? How many steps does it take?

Eric Drexler knows perfectly well what an enormous undertaking he's urging upon us. So, of course, did Arthur C. Clarke and his egghead buddies just after the Second World War when they tried to sell an incredulous society on a package of ridiculous ideas, each more laughable than the one before it: satellites in earth orbit, telecommunications bounced off them, rockets with human pilots, trips to . . . get *outta* here—the *Moon*!

And Drexler is conscious, it's plain, that working engineers might take umbrage at his bold vision, his skidding across details:

> Many of the steps described will, if attempted, spawn a host of subproblems, each demanding long, hard, and creative work. It would, however, become burdensome to point this out at every turn. Developments that will one day make molecular manufacturing fast and easy will result from efforts that are slow and difficult.

Certainly he does not expect to reach the grail, the Genie, in anything like a single heroic leap. Drexler, according to *Scientific American*'s Gary Stix, is Mr. Peabody; cautious, pedantic to a fault (despite the wild stuff he might share with Ralph Merkle and his colleagues in the back room). You move forward from atomic force microscopes and scanning tunneling microscopes by a series of deliberate steps, mapped by *backward chaining*. That is an analytical tool leading an engineer step by understandable step in reverse, from a desired end to a possible starting place. With every step backward you map from what's required to that prior sensible step (yes, if you want D, it helps to get there from C), and then back another judicious step (and B would get you to C), all the way to what you can do already, or nearly so (ah, we could reach B with the help of this big, clumsy Atomic Force Microscope here on the bench).

Four steps to nano minting

Drexler provides a chart of four likely steps or stages in mechanosynthetic assembly. Even doing this sketch work, he is careful to note that Merkle, for instance, has developed a quite different pathway in which "the most likely subsequent steps would differ greatly" from his own program. That's okay. If there are many potential early steps, backward chaining is the way to go, multiplying possible paths for exploration.

First-stage nano—the earliest steps

The primitive first stage builds its gadgets in a solution of chemicals, a wet medium in which reagents—the chemicals you use to prompt and control reactions—are transported more or less at random by diffusion, the way a drop of dye spreads, threadlike, in water.

It will be shockingly inefficient, and applied forces will be small. Cycle times required to build one thing out of another will be about a second. The things being built will be (like enzymes) folded polymers—long strings of linked molecules—with ten to a hundred parts.

At this point in mint evolution, instructions will come largely from outside by acoustic pressure waves, translated by simple decoders far less complex than the onboard computers of later generations. The building blocks, each made from perhaps fifty small strings of molecules, will self-assemble, tossed together by the ceaseless Brownian motions of their suspension medium. But the process is by no means random. "If building blocks are added to solution sequentially (and removed afterward), each will have a unique destination so long as the assembly sequence exposes only one interface of each kind at any one time."

Alternatively, simple minting components can be put together by AFM-directed synthesis. And that takes heaps of knowledge and control, as Drexler admits candidly: "Image interpretation software able to determine the types, positions, and orientations of specific sets of reagent binding molecules; control software to automate reagent positioning and sensing, thereby automating the execution of long sequences of reactions"—you get the picture. And all of this, he wrote confidently, can be forecast as feasible within "a fraction of a decade."

Well yes, but that was 1992, and years later Ed Regis—author of the very positive *Nano!*—was complaining in the *sci.nanotech* newsgroup: "Why don't we have even one STM chemically joining one molecule to another molecule, thereby offering a proof of concept of mechanochemistry? The only time this reaction has ever been tried (by Don Eigler, of IBM Almaden), it failed."

At least one can see that such elementary assemblers, even if they still have not been built, don't call for the near-human artificial intelligence controllers that Augustine thinks would simply lie down petulantly and refuse to work for us.

Mob rule

Another expert, Martin Krummenacker, notes: "This problem of putting together a self-contained system from a cruder level of sophistication by hooking up a number of more primitive parts is often encountered in computer science, and is termed 'bootstrapping.' "[8] Krummenacker has detailed the kinds of lattice-arrayed molecular building blocks (MBBs, which one might pronounce "mobs" due to their great number and anonymity). "To assist the human designer," he remarks, "computer-aided design tools are necessary that are able to autonomously generate proposals for polymers that fold into a tightly packed structure."

His own program blindly grows polymers to fill lattice cavities that otherwise would leave the mobs with too much conformational freedom, so they'd fold up into useless shapes. Still, within such programs there remain points "where one could add in smarter and higher-level strategies than the random selection mechanism used now." For the moment, those smarter strategies will come from keen human minds. Genuine human-level AI might have to wait for mature stage-four nano (to be outlined shortly), the appropriate technology for building brains, while the full development of truly mature, awesome nanotech will depend in a, well, a *bootstrapping* way upon human and superhuman versions of AI . . .

Second-stage nano

The humbler second stage of Drexlerian mint bootstrapping shrinks to submicron scales. Now, rather than depending on self-folding polymers, we'll see cross-linked polymers with a cycle time of perhaps a hundred-millionths of a second, a considerable speedup. Energy efficiency remains low, and the medium used is pure liquid. Feedstocks, arrays of complex molecules, have to be prepared in advance. A mint of this marque might take an hour to copy itself.

Third-stage nano

By Drexler's third stage, instructions are still being delivered from outside the fabrication containment site, but cycles times are down to ten microseconds and "systems are built with internal control and data storage devices, [so] brief instructions can activate complex subroutines, greatly relaxing constraints associated with data transmission rates." Hence, mints develop flexibility, able to assemble a variety of different modules or molecular building blocks.

The fourth stage – mature nano

The fourth stage (finally) is mature minting, using assemblers made of up to a billion molecules. It builds with solid diamond-oids, put together with precise positional control. Because the reagents in use are extremely reactive, the workplace is now a high-grade vacuum. Feedstocks are simple molecules, fetched in and sorted by nanoscale conveyor systems. Cycle times have shrunk to a millionth of a second, while error rates, which have been improving steadily with the degree of positional accuracy, are down to one misaligned atom in a thousand trillion. Held in place, guided firmly with considerable force, each atom is bonded in the exact place set out for it by the considerable power of advanced computational modeling.

None of these projections, however, seems to call for machine intelligence at the human scale. Of course, it's perfectly possible that with the bounding leaps in AI research and construction made possible by stage-four minting, we'll see the emergence of cheap machines capable of not a trillion calculations per second (a teraflop, already surpassed by supercomputers) but a thousand trillion (a petaflop) or better. As a kind of free side effect, such astounding devices will have little trouble meeting the Turing test—passing for human, that is, in conversation. The two technologies will then climb, as it were, upon each other's shoulders.

That would bring the Spike to us in earnest, and faster than we might have expected.

Then again, perhaps not, if Augustine's disillusionment with the nano prospectus and its ancillary breakthroughs (AI, cryonic resurrection, uploading of human minds into digital cyberspace, all that) is anything to go by. Here is his fairly scathing assessment, aimed at demolishing easy acceptance of Drexler's delightful and scary outlook:

> Eric has taken the ideas of apocalypse, resurrection, and eternal life—memes which propagate themselves like dandelions in your lawn—and put them into a plausible scenario, so for the first time it seems they can actually, physically happen.
>
> According to Eric, the Assembler Breakthrough is an inevitable fact looming ahead of us. Resurrection has become an actual possibility; and not only a possibility. But before we reach the resurrection, we have to pass through the Dangers of Nanotechnology, which could destroy the world. All

the elements of the Apocalypse are there, transposed into a form which makes sense to a modern person who believes computers can do anything.

Something new under the sun

Meanwhile, of course, Augustine's former allies stared in stark disbelief. Didn't he understand *anything*? Mints are *not*, let's say, the miniaturized equivalent of the Chinese economy, pumping out cheap imports of shoes and spanners and old stockpiled copies of *The Thoughts of Chairman Mao*. Nanofacture is *different in kind* from all previous economies.

Why? Because now, abruptly, it's the *means of production themselves* that are replicable. And from elements cheaply— or even freely—available in the environment, wherever you happen to be.

It might cost a zillion dollars and exhaust the mental reserves of an entire generation's finest Coke-fueled minds to build the first stage-four mint, but once it's there in its vacuum tank, once its specs are in the can, you can tell the thing to fission and multiply, and it will make its twin, and they'll make another two, so you have four, then eight, then sixteen ... and by god at the end of the day you will look into your garden, at your handiwork, and you will see that it is good.

Hacker, code my steak

True, you still have to inform this seething pond of eager beavers how to make a pair of nylon stockings and a slab of sirloin, because all it has in its tiny distributed artificial brain is a vast recipe for making itself. But surely that's the *easy* part. Plenty of people will *love* to do it for fun.

Once you have a working matter compiler, you can share the job of writing the programs to ten or a hundred thousand nano hackers on the Internet (since, by then, the Web will be ubiquitous in the developed world). Today, computer enthusiasts write reams of code and post it for free, or as shareware—for the plea-sure of it, for the art, to show how good they are, for peer review and correction, to share useful code patches (repairs), to earn the esteem of their peers. Increasingly, vast code structures like the operating system Linux (widely regarded as a superior and cheaper competitor to Microsoft's Windows operating system) and the Apache server are "open source." That is, their code is freely available for elaboration by a host of enthusiasts who par-ticipate in their development and evolution, supported and cor-ralled by companies (such as Red Hat, SUSE, and Caldera for Linux documentation, authoritative debugging, etc.) who charge for their services rather than for code or hardware.[9]

This new path to product creation and unorthodox marketing, highlighted when Netscape suddenly released its source code in January 1998 in a bid to save itself from being strangled alive by Microsoft's Internet Explorer, are discussed exhaustively and en-tertainingly in celebrated documents, freely readable on the Web, by Eric S. Raymond, president of the Open Source Initiative. "The Cathedral and the Bazaar," "Homesteading the Noosphere," and "The Magic Cauldron: Indistinguishable from Magic" argue, "When your development mode is rapidly iterative, development and enhancement may become special cases of debugging—fixing 'bugs of omission' in the original capabilities or concept of the software. Even at a higher level of design, it can be very valuable to have the thinking of lots of co-developers random-walking through the design space near your product."[10] Those codevel-opers are just unpaid *users* of the growing product, freely con-tributing to the practical and aesthetic task of making the code elegant, simple, and . . . simply the best. (In some cases, however, programming is done by employees of an open-source software provider such as Red Hat.) We shall return to these arguments below, in discussing the real-world economics of developing as-sembler technology.

A slightly different example, perhaps closer to a hardware na-nomint than to purely software packages: MP3 digital music is increasingly available on the Web, and while the record compa-nies were nervous at first, they are adapting, learning to make money even when music of every kind is running free.[11] In the future their equivalents will be up all night coding ever more subtle and cunning Killer Apps for the mint, selling them from garage start-ups (until the code pirates sell the CD-ROMs at a 99 percent discount, or generously steal and repost your code on the Net), or finding street glory, like the graffiti tagger kids did back in the twentieth century, by sending their own applications out instantly onto the Net as shareware and freeware.

Reality check—hard economics

We need to pause here and take a breath. Who is supposed to be funding this leap into future technology, especially when its end result is a new kind of productivity that bites the hand that feeds it? Grant for the sake of argument that Drexler's trajectory of development is not impossible, even that it makes quite a lot of sense. Somebody starts with first-stage molecular nanotechnology in the lab, and step by step a new industry emerges, culminating in fourth-stage or mature minting. Each of those steps needs to have its independent economic rationale. Corporations and private investors must be convinced that their money will earn a good return—a better return than if it were put into hamburger franchises or real estate. Yet how can they sensibly arrive at that conclusion if the extravagant, utopian dreams of nano's more vocal enthusiasts have any validity?

But then technology really *has already* managed to produce more and more for the same amount of effort, lowering the prices of plenty of goods. Politician and futurist Barry Jones saw how things were going in the microelectronics end of the economy, and in *Sleepers, Wake!* put his finger on what will also drive the development of minting, only much more so. Miniaturization, he notes:

> permits an exponential rise in output together with an exponential fall in total inputs—energy, labor, capital, space and time. In economic history there is no remote equivalent to this . . . This overturns the folk-wisdom that for every ad-

vantage there is a corresponding disadvantage or price to be
paid ("You can't have your cake and eat it too").[12]

Cybernetics theorist of the 1950s Norbert Wiener foresaw
computer-controlled automation as a kind of equivalent to slave
labor, and hence supposed that human workers in competition
with these machines would be driven into ruinous slave wages
themselves. That hasn't happened, by and large, although there
is an increasingly large number of people either without work or
sequestered in part-time work for McDonald's or the local hy-
permart. Still, as Jones points out, Wiener's fears were misplaced
because he "failed to read economic history . . . Slaves don't provide markets for sophisticated, diverse, personalized production.
Computers break down hierarchical work structures, providing
the challenge/opportunity of individual creativity."[13] Even so, he
acknowledges, there are victims: those lacking literacy and nu-
meracy, people simply not able to join the "information rich."

Still, all this is so only because investors can make money from
information technology and the hardware that supports it. The
technical dynamic that results in the frequent doublings recorded
by Moore's law might falter if consumers suspect (as many do
now) that they are paying for "bloatware," massively redundant
code accumulated to service all conceivable users on every known
platform, requiring colossal extra amounts of computing grunt
to do pretty much the same jobs about as well. If that corrosion
of confidence continues, the national and international markets
for an ever-growing number of small computers and their software will become exhausted. Meanwhile, the cost of refitting fabrication plants to create chips at smaller and denser scales
balloons (although these plants will be refitted for advanced uses,
amortizing more of the initial investment). Why should canny
investors choose to move their money into the nano field if they
can see, at the end of the development tunnel, an even more
frightening prospect: universal or general assembler machines
that literally compile material objects, including more of themselves (given, perhaps, the right "vitamins" or rare trace elements
to restrict utter runaway self-replication)? Where's the profit in

that? Who will invest heavily in developing a magic mill that makes copies of itself almost for free?

Brutally: who pays, and why?

The taxpayer will pay

One answer, as usual, is that government-funded as well as privately endowed university researchers continue to pump out endless astonishing discoveries and applications—that category known as "pure" or "disinterested" research. A perhaps-unexpected instance has been NASA's Nanotech Team, part of their Science and Technology Group.[14] Their interest should not be startling (except for the clear vision displayed, which reportedly is now faltering). As atomically precise manipulation of matter grows increasingly common in labs, they note, the implications for superstrong, lightweight aerospace materials and advanced small computers are unavoidable. The time line to full molecular nanotechnology might be measured in decades, they admit, and nobody yet knows the precise path: "diamondoid, fullerene, self-assembly, biomolecular, etc." The team—accredited scientists such as Al Globus, Jie Han, Richard Jaffe, and Glenn Deardorff, of NASA's Ames Research Center, Mountain View, California—listed these remarkable goals: "Develop Ames' computational molecular nanotechnology capabilities; Design and computationally test atomically precise electronic, mechanical and other components; Work with experimentalists to advance physical capabilities."

A 1997 technical paper by members of the team, in the new journal *Nanotechnology*, described how molecular-sized gears had been simulated by a NASA supercomputer.[15] In the simulation, benzyne molecules (C_6H_4) are attached to the outside of a nanotube, forming gear teeth cooled by helium and neon gas flows. The molecular gears are electrically charged by a laser beam, spinning the tiny gears at six trillion rotations per minute.[16] The team explicitly targeted its work, ultimately, at nano assembly of large-scale products. Creon Levit, of NASA's Numerical Aerodynamic Simulation Systems Division, stated: "A computer program would specify an arrangement of atoms, and the matter compiler would arrange the atoms from the raw materials to make a macro-scale machine or parts." Nor was Dr. Globus shy: "We would like to write computer programs that would enable assembler/replicators to make aerospace materials, parts and machines in atomic detail."

In a remarkable and comprehensive paper published in the *Journal of the British Interplanetary Society* in 1998, the Ames team, with the inclusion of Ralph Merkle, made it plain that their guiding light was none other than Eric Drexler and the design-ahead analyses from his Foresight Institute (of which Merkle was, and is, an important member).[17] Here was a source of ready research funds and talent providing, if only in a small exploratory way, some sorely needed nano innovations, bridging the chasm between our current rudimentary knowledge and a working matter assembler. NASA was clearly serious about this stuff: a four-day January 2000 NASA-sponsored conference in Houston, Texas, on "Nano/Microtechnology for Space Applications" deployed a small army of invited speakers and more than a hundred technology presentations.

Another important team associated with NASA is the Caltech Computational Nanotechnology Project, under principal investigators William A. Goddard III and Dr. Tahir Cagin, both of the Materials Simulation Center. The team's declared objectives are to use large-scale molecular dynamics to study and simulate the dynamics of nanomachines and molecular assemblers, use quantum mechanics to study and develop diamond mechanosynthesis tools, and to employ well-known statistical simulation techniques to examine self-assembled monolayer strategies for constructing components of nanoscale systems. "Ultimately," they say, "we need a programmable synthetic system to make a real device. Even though we may not have tools for all the chemical steps and may not have designs for all the pumps, and engines, and transmissions needed, we propose to study the dynamics of simplified prototype assemblers."[18]

Private investors will pay

That is the public or academy-subsidized approach. Another answer is that the market does, after all, fund investments evaluated by a variety of horizons, some near, some farther off, some totally "blue sky." IBM has been profligate in funding pure research, and has reaped rich rewards, although maybe this is changing.[19] Corporations such as Geron (in cancer, stem cell, and telomere research) and Zyvex and Ntech (in micro- and nanotech realms) work on both immediate and long-term studies into topics that might yield truly flabbergasting outcomes eventually and in the meantime could bring more limited products to market.

Ultimately, feedstocks might be obtained in situ and gadgetry made so small or ubiquitous (Ufog, or "utility fog," which we shall investigate shortly, for example) that not much bulk needs to be shipped. But that's the longer term, and the road to such Santa Claus mints is not self-evident. An articulate Chicago electronics and system engineer, John S. Novak III, argues that "if we reduce physical production to the level of software compilation (which is what a Santa Claus machine would do) we still cannot necessarily claim that the open source movement will be the salvation of mankind in every respect." That's because teaching the mint how to compile a product will take detailed, specialized one-off programming. This isn't just a software constraint—as an RF/microwave engineer, Novak knows how much the devil is in the fine-grain, real-world detail. Open source or not, who'll be willing, without pay, to do the unglamorous scut work? "Very few people are going to *want* to write the software for something as boring as your automatic timed Toastermatic 2010."

Every bozo knows what a decent chair should look like, even a good office chair or recliner. If not, it's a simple matter to look up ergonomic studies and implement them. Does every bozo know what a good television should look like, on the inside? Well, no, probably not. Will every bozo know

what is required to make a nanobot that will reverse whatever foul age-related biochemical reaction is thinning my
hair?[20]

Good practical sense. It assumes, you'll note, an early stage of
nanoassembly and AI. But the advent of fast semithinking machines might allow nifty, cheap reverse engineering, carried out
by systems that are not intrinsically expert. What's more, at some
point—if the Foresight Institute's vision is realized—your com-
munal or private nano-soup tank will decompose a Bugatti or a
banjo and store the algorithms on the Net for free downloading.
Nobody expects relentless rugged individualism, roll-your-own
everything. You'll trade your trichological research into hair replacement genes for my new nano back-scratcher or kangaroo
roast hot from the assembler. (Actually, we already have a market
to handle such exchanges, soon to be fully accessible on the Net,
so reinventing the commonweal might not be necessary.)

Novak does not deny the lure and satisfactions of hacking elegant, gratifying solutions to difficult problems. A generation of
fairly wealthy programmers and engineers, he notes, might well
take early retirement in 2015, turning their skills as "a loose fed-
eration of yeoman tinkerers" or hobbyists to AI and nano compilers, among other challenges. But at least in the early decades
this will only occur in a larger commercial environment where
corporations continue to squeeze profit from technical advances.

We need a path to widespread nano that's equal to the path
from early spreadsheets to Microsoft Office software. That remarkably flexible set of applications can be purchased fairly inexpensively, or (illegally) stolen and burned or copied to
multigigabyte drive in seconds. In the meantime, ancillary industries will still thrive—transport, etc.—but those are not the
interests needed to *invest* in making nano happen. This looks like
the most reasonable path: build a clunky thing that does something unique and useful, then a better small thing, then a wicked
cool smaller thing, and so on, and sell them sequentially for the
benefits they bring to existing productivity. And have enough

money in your swag at startup that you can afford some mistakes and missed targets.

A case in point—Zyvex

Consider Zyvex, a developmental engineering and directed research startup company founded in April 1997 in the Telecom Corridor of Richardson, Texas.[21] Its explicitly declared goal is "to build one of the key pieces of molecular nanotechnology; the assembler." They acknowledge an intellectual debt to Drexler in forming this goal (while recommending close attention to such refereed journals as *Nanotechnology, Surface Science, Journal of Vacuum Science and Technology,* and *Journal of Physical Chemistry*), and state that "our first product will be some type of assembler, which others can buy and set to work to make useful things for the market they are already familiar with. We don't expect to go conquer all known markets and make all possible products by ourselves. The assembler will be enough for starters." The company was not seeking outside funds in early 2000, and announces "a 5–10 year time horizon for its first revenue."

What kind of assembler do they have in mind? Certainly not a handheld miracle machine, or an AI-controlled Genie. Rather, they envisage:

> a system of unspecified size (possibly quite large initially— say one by two by two meters), capable of manufacturing materials or arbitrary structures with atomic precision, getting nearly every atom in the desired place. An error rate of 1 in 10^6 would allow meaningful structures to be built, and is probably a reasonable goal for the first generation. A more mature system would have error rates in the 1 in 10^{12} range or better.

A Zyvex assembler might be regarded as a child of a current Atomic Force Microscope (or Probe), powered and controlled from the outside, but ideally able to copy itself and even produce an improved version (again, naturally, following instructions provided by engineers). To handle useful amounts of raw materials and produce salable quantities of goods, huge numbers of assembler manipulators will be governed by a central control processor. Zyvex expects to build assemblers with a million positional devices, each doing one thousand tiny tasks a second. That's a billion molecular building blocks put into exact place every second, for each chunky assembler—still not a great deal in visible terms. The second generation might graduate up to 10^{15} building units a second (that's a million billion), and the third generation commercial assembler would process 10^{24} molecular Legos a second.

Their game plan for reaching even the earliest of these mints is cautious. They'll start with current microelectromechanical (MEM) technology. A MEM assembler has the advantage, as a testbed, that what it's compiling will actually be visible under a scanning or even an optical microscope. Next might follow a fullerene or nanotube assembler, handling and chemically bonding pregrown microscopic carbon tubes. Trust me, it will not be building a diamond spaceship out of old potato peelings, however. This humble forerunner of the mature mint will put together very strong structural beams, welding carbon tubes into rods or pipes stronger and lighter than steel. Zyvex has in mind a device that is far from perfect but can function without the perfect vacuum or difficult temperatures needed in more classy designs. So it won't build diamond or diamondoid, but will make useful products, and compile them cheaply. "All we need," Zyvex's promotional Web site confesses, "is a suitable reaction, a suitable catalyst, a suitable positioning device, and some luck getting the catalyst stuck onto the end of the positioning device just right. Does that sound hard?" They are honest, and wry. "Yes."

Even so, by 1999, the company had already announced the launch of Zybot Mark 1, a MEM micromanipulator with two little grabbers able to move with five degrees of freedom, "holding MEMS Precision Instruments tweezers." The Zybot will make more elaborate MEM gadgets, and so on—all the way down, perhaps.[22]

So you can start to see how the program comes together. Engineering research like this need not make a profit for up to a decade, but at the end of that period it reasonably hopes and expects to bring out product that will sell to the targeted industries: chemical, engineering, building. Meanwhile, they are gathering the hard-won trial-and-error information needed to build a true Drexlerian mint. Another company working the same vein is Ntech Corporation, which as part of its push to a practical nanotech are already making and shipping a scanning machine (it looks like a standard flat scanner) that can detect and read "StuffDust" microtaggants digitally encoded with readable data. These smart dust motes can be coded with binary-format information and sprinkled into library books, say, or into hard-copy files in a registry, then read out in translated form for identification or for stocktaking purposes.[23] (Thomas McCarthy, Ntech's president, is not content to help create the foundations for a future nanotechnology—as we shall see in chapter 7, he has worked through many of the implications, for industry, war, and peace, of various implementations of minting. His conclusions are clear-eyed and quite frightening.) Ntech has developed a range of sensor products for the shipping industry, monitoring vibration, temperature, humidity, and location using micromachining able, they claim, "to take advantage of nanoscale phenomena to do their job." Again, this is MEM-type technology, but creating systems working at this scale provides the foundation for nano—and pays the bills while that longer-term research is in train.

Recall the development curve of today's most ubiquitous personal computers, which started half a century ago as immense clunky behemoths, passed through phases or epochs where prices dropped by consecutive orders of magnitude so that first governments and their militaries could own computers, then huge businesses, then smaller businesses with the arrival of spreadsheets, then secretaries and students as word processors became common, and now the gadgets are getting as routine as personal phones (an even more explosive marketing phenomenon), with specialized variants lurking under the hood of your car or run-

ning your microwave cooker. It will be the same with nanotechnology. Nobody expects Kmart assemblers to arrive overnight. By
the time they do, a lot of money will have been spent, a lot more
recouped, and the economics of the world will have transitioned,
by small and large jolts, into a new equilibrium (if we are lucky)
as odd to our eyes today as the world of the Keynesian welfare
state would have been to an eighteenth-century merchant.

Even experts who differ with Drexler over the wilder outcomes
of the new miniaturization technology agree that nano spells immense change in the economy. Professor George M. Whitesides,
Mallinckrodt Professor of Chemistry at Harvard, sees nano as an
extension of MEMs. He notes, as we saw above, that shrinking
the scale of semiconductors, say, pushes up the costs of fabrication plants. "If you want 20 percent return on investment, and
you put in $10 billion, how many microwidgets do you have to
sell every year for the few years that that fab is the state of the
art? The answer is, a lot. And people who have to put up the
money don't like that."[24] In the immediate future, Whitesides
expects impacts from inexpensive microtechnology rather than
nano. For example, he predicts that instead of a newspaper "you
might buy a sheet of paper; the back side of it would be a battery,
the front side of it would be a display. You read it, scroll to find
reference works on it, see animated illustrations, and when you're
done, you throw it away." But marketing common consumables
like reusable or disposable and downloadable paper with embedded microelectronics (these already exist, but are not yet cheap)
will provide some of the industrial base for more adventurous
miniaturizations. Whitesides warns that you can't just shrink a
machine and expect it to work. Those planning the creation of
mints via a developmental trajectory like Drexler's four stages will
take that for granted. It will be necessary to fund each step of
the molecular nano revolution by careful, incremental advances
that are each profitable but allow the shift to the next stage as it
emerges from the lab.

When that occurs, though, we really should expect to see some
extraordinary leaps. Some of the limits common to today's economics of production and distribution (but not all, by any

means) will blow away. There is good news and bad in this prognosis.

Nano and the limits to growth

Dr. Rich Artym, a British software engineer, sees the shift from manufacturing to nanofacturing as "an utterly colossal change to the effective limits to growth today." Once rudimentary nanoware is out there, he claims, the individual suddenly has astonishing power even without artificial intelligence systems, let alone SuperIntelligent machines. His scenario suggests just how dramatic the changes might be:

> if a person can program a nanosystem to create (say) a saltwater pipeline which extends itself a short distance into the rock under his property, he or she can in principle create a pipeline extending into the nearest ocean, *without* requiring any extra capital or manpower or any other resources. Such power in the hands of everyman completely destroys the basis of our current economy, which is founded on the maxim that it is more or less impossible for any single person to do anything major in a material sense by him/herself, needing to *purchase* everything, which keeps the economy rolling and allows governments to reel in taxes.

What's hardest in this projection comes right at the outset: bootstrapping from bulk technology to the entirely different economics of minting. Once that step's been taken, conservative expectations fly out the window.

Mocked though they are by the staid for their imaginative extravagance, Drexler and his colleagues strive for PR motives to remain "within application areas that the *status quo* will not find too fantastic to believe," Artym notes, "but the reality of the matter is that their examples are only the beginning, and are all exceedingly tame."

Well, maybe, but go back a step and think this through. Undermine the landscape with *saltwater pipes*? No planning permits, no environmental impact statements, just *do it*. Send your nanomole burrowing a mile deep and then a sharp turn right until it hits the sea, building as it goes a sufficiently strong diamondoid-shelled capillary with, let's say, microscopic cilia every few centimeters to stroke the water molecules uphill back to your tank, gnawing its way through rock and telecommunications cables and oil wells and other people's hose lines—

No, that won't work. So make the mole gadget at the growing edge *smarter* than that, able to sense what's up ahead (via acoustic pulses, echo detectors, electric field monitors, whatever does the job), able to work around anything fabricated that it finds in its path. This sounds suspiciously like the autonomous AI that Augustine the skeptic says will be a requisite, but maybe such exigencies can be listed in a prepared look-up file. We could do that already, perhaps.

Onward it chews, raising its tiny snout at last from the gooey ocean floor, or perhaps from the bank of a local stream (to everyone else's annoyance, when the stream suddenly dries out in the hot sun), sucking back water, passing it through the layers of filters it has nanofactured on its journey. Clean water runs up one set of capillaries, and solutes—trace elements of every kind, the useful and the toxic—are segregated, bound and disposed of into adjacent rock if undesirable, otherwise selected and concentrated and sent uphill to your storage tanks.

And all this furtive, sinuous digging and sorting and pumping starts occurring everywhere in the planet, each set of waterpipes coded to one of the six billion, the ten billion, the twenty billion humans on earth, a world suddenly honeycombed by wormholes—

You have to hope that this greedy maze of feeders doesn't trigger an earthquake, or simply hollow out the ground under your feet so that abruptly you tumble out of your green paradise into a hellish pit. But such alarmism might be ill-considered, given the modest dimensions of these free mint-made supply routes, and the very great volumetric spaces that stretch downward beneath the skin of the world. And perhaps it isn't too tricky to instruct your hydrophilic mole to pass up all the scenic rivers and ponds and hydroelectric dams and strike instead, in its mindless pioneering way, for the distant coast.

The Overtool

The nanomole, and its construction brethren back at home matter-compiling your new winter wardrobe from the on-line fashion catalogue, assembling a rack of lamb with all the trimmings for dinner, and paving the new roof of the new study with new photovoltaic cells, might be a composite of Forrest Bishop's modular universal assemblers or "Overtools."

Bishop's scheme is that a bunch of these handy all-purpose units would have enough grabbing power, positioning dexterity, and so forth, to make up the core of a *general assembler* with a manipulator arm. This gadget is made of smaller cubical Active Cells sliding in one of three directions past each other, capable of macroscopic assembly as well as diamondoid mechanosynthesis, not forgetting self-replication.

What's the minimum needed for such a universal tool? "Enough variability in its permitted degrees of freedom to perform the immense number of different tasks needed," plus "a reliable algorithm for the assembly process." Add an "XYZ gantry" to an aggregate of simple cells, holding at the tip of its mechanical arm a number of specialized tools, and you get a gaggle of gantries "holding, straining, and rotating an arbitrary workpiece, while others fetch reactive species or perform abstraction reactions."

With a modular design like this, you can scale up or down, from the nanoscale to the near-astronomical: a Bishop "strained shell" design could be a mesoscopic "1700 nanometers, or perhaps 1700 kilometers (space-based). It would be necessary to build such a large structure of something very strong and cheap, like diamond." It's a measure of our progress through this unusual landscape that a phrase like that about the cheapness of diamond should not have raised an eyebrow.

Key questions remain: how autonomous and smart do these minting mechanisms need to be, if we're to treat them as part of the household furniture, like a home computer or a video player or a microwave oven? And is it true that copious software will swiftly become available, cheap or free, so we can use the raw power of replicating Overtools and mobs and nanites, call them what you will, to do our bidding?

Would *you* work without pay?

Actually, as we've seen, people *don't* have to be paid to work, if it's their idea of fun.

That's one surprising thing the culture of abundance has proved, I think, especially among the young and those old enough and secure enough to be able to play with computers in their spare time. It's the same with physical exercise: excruciating exertions in the gym will build you a better, automorphed body, and you'll pay through the nose for the privilege of toiling like a workhorse.

Kids who refuse to do their homework even when bribed will spend hours, days, months, cramming the higher centers of their brains with arcane and subtle rules for the latest hideous computerized combat-cum-strategy game. Many of the great breakthroughs in commercial programming have emerged from brilliant obsessed kids in garages (called "nerds" by less gifted, baffled, secretly envious onlookers). Even if the ultimate lure was wealth beyond the dreams of anyone except Bill Gates, still, the prod was sheer intellectual vanity, the aesthetic yearning for elegance and concision and mastery.

But oughtn't people be motivated only by the quest to create profit through diligent work? *Decent* people, that is—not the wicked and work-shy who, some suppose, skulk about planning to rip off their hard-won wealth. The ethic of selfishness is widely praised.

In many of the people imaginative enough to envisage nanotechnology, machine intelligence, and the Spike itself, I've found, there is a curious blind spot. Perhaps it's because you need to be a rugged individualist to break free of dogma, or perhaps it's because many of these thought experiments are spun in the United States, where vast inequities of wealth are associated with resentment, crime, self-destructive drug abuse, self-inflicted stupidity . . .

I find it interesting that this background assumption—that many other citizens are malevolent, lazy, stupid (or cunning)—tends to be more prevalent in societies where extremes of capitalism *or* state socialism have taken hold. In many social democratic European nations, and Canada and Australia, where a shifting blend of these approaches operates, there has been less street violence, far less polarization of rich and poor and hence of envy and fear (although that is changing, and I'm disregarding the usual brutal treatment of indigenes in the latter two nations).

In general, it seems to me, people will help each other, when asked, unless there is an overwhelming motive for not doing so. Bizarrely enough, given my enthusiasm for computers, I'm pretty much of a klutz with the things—and would have had a great deal more trouble accessing the Internet, for example, were it not for various people who simply chose to be generous and help me out. In return, but not with any strict accounting being recorded, and certainly not out of any saintly generosity, I do what I can to help other people in areas that pertain to my skills. Don't you?

Potlatch beats greed

In some societies this sort of behavior is formalized as *potlatching*, a kind of wild partying, a ceremonial blow-out that preindustrial cultures, such as the southern Kwakiutl Native Americans, once employed to even things up when fortune favored one or more families over the rest. Of course, the feasting and rich hospitality paid off in social status and respect.

As less and less jobs remain in a rapidly approaching future modified by AI and nanotech, potlatching seems to me a possible pattern for decent human relationships. We might not always be able to love each other, but we certainly won't get far if we assume that everyone else is out to grab our hard-won goodies.

Still, I have to admit that, in my own country and elsewhere, consumers have lately achieved a far more pleasant standard of living than ever before by importing huge amounts of very cheap consumables from the sweatshops of South Asia. Doing so has effectively wiped out the small domestic industries (except those illegally running their own sweatshops) that used to produce clothing, light electrical goods, etc. In the long run, of course, nanofacturing will destroy the old industries anyway, on a mammoth scale—but with any luck (*pace* Augustine) the goodies flowing from the cornucopia machine will be *so* cheap that everyone will gain by their existence.

In the meantime, we can learn from writer Bruce Sterling's Viridian Green Manifesto, a techno-friendly, techno-embracing call to arms for the twenty-first century declaring, "The aggressive counter-action to commodity totalitarianism is to give things away. Not other people's property—that would be, sad to say, 'piracy'—but the products of your own imagination, your own creative effort."

Right now, the inner-city industrial suburb where I live is full of empty factories. Thirty percent of the low-scoring kids who leave school early, the ones who once would have had a fairly nasty life working in those emptied factories, now cool their heels at home or on the corner all day long. They are supported by the dole but chafe for a meaningful life. But adult consumers (at least half of them, since voting is compulsory and therefore universal in Australia) keep returning politicians who ordain the "level playing field" doctrines responsible for the ruin of their neighbors. It's the very reverse of potlatching. The lid is clamped down on the pot, and the latch is locked.

At any rate, even if the first nanites prove not to be independently intelligent, contriving their own software upon request, still, the top 5 or 10 percent of the minds of any culture with food enough not to wake hungry in the night certainly *are* that intelligent, and ready to use the finest brains on the planet to code up the machine *du jour*. If that machine is a mint, the replicator algorithms will be spilling onto the Web like jewels (and mouse droppings) tumbling from some sack discovered in Aladdin's cavern.

Maybe.

Or maybe it will be much, much harder than that. Absurdly, deplorably harder. So hard that it would make you weep with frustration.

Utility Fog—nanos everywhere

Suppose technology does take the paths predicted by the optimistic. Nano could become as commonplace as the air you breathe. Taken literally (nano aloft in the air), that's the hope of JoSH, J. Storrs Hall, moderator since 1988 of the Usenet *sci.nanotech* newsgroup. He calls his neat idea the "Utility Fog," which is just what it is: a permeable fog of floating nanites drifting across the face of the planet, like the flying microscopic hive machines in Stanislaw Lem's classic 1964 Polish SF novel *The Invisible*, ready to lock together in local clumps like Forrest Bishop's Active Cells and . . . do your bidding.

It's a prospect to muffle you, perhaps, in claustrophobic anxiety. The sweet air is gone—and the polluted, friendly city air, for that matter, reeking of car exhausts and restaurant odors, people smells and the flushed colors of evil chemicals tinting the edges of the afternoon sky like a Turner painting—replaced by a choking haze of tiny floating nano foglets, jostled by Brownian motion, trillions upon quadrillions of them as far as the eye can see.

You can relax now. That's not what JoSH has in mind at all.

His ambition is sweeter, far more congenial, and, well, *poetic:* "What I want to be when I grow up, is a cloud."

J. Storrs Hall, that is, doesn't just wish to live in Fog City, a metropolis at the edge of dream, built from buoyant, instructable nano foam. No, he wants to *be* a nano cloud, his mind and memory and volition uploaded into an adaptable robotic body minted of smart dust.

Utility Fog need not be colonized by uploaded minds, not to

begin with. Just by itself, it's *intelligent stuff*. It's kin to the Drex-lerian smart paint we met in chapter 3, but dispersed in an im-palpable spray through the room and maybe out into the street. And of course it's only impalpable while it's in its free-range state.

Give Utility Fog the correct signal, or allow its onboard pro-cessors to estimate the needs of the flesh-and-blood people wan-dering through it, and each minuscule aluminium oxide–shelled foglet can reach out some of its twelve arms and catch the arms of its neighbors, jostle, shake down like a row of marines calling off, create a lattice in depth, add detail as a macrostructure emerges from computer memory into literal reality, becoming— *anything*, pretty much, within reason. A chair for you to sit on, or just an invisible support as you loll back into the air. A hat for your head to keep the rain off, or an entire house with walls and roof to do the same thing, designed according to Georgian principles down to the slightest elegant detail. You can't eat Utility Fog, though, or rather you *can* but it won't nourish you. In fact you can breathe it in and out without doing yourself any harm, just as you do now with all manner of other suspensions at viral and even bacterial sizes.

> The Utility Fog operates in two modes: First, the "naive" mode where the robots act much like cells, and each robot occupies a particular position and does a particular function in a given object. The second, or "Fog" mode, has the robots acting more like the pixels on a TV screen. The object is then formed of a pattern of robots, which vary their prop-erties according to which part of the object they are repre-senting at the time. An object can then move across a cloud of robots without the individual robots moving, just as the pixels on a CRT remain stationary while pictures move around on the screen.

This is an immensely appealing conceit, once you have it visu-alized. You could "hold" a ball made of fog and "throw" it high across the room, watching it soar in its parabola and crash splin-teringly into the mirror . . . without anything physically passing from your hand to the mirror. An *image* of the ball is all that spins and crashes, calculated to a nicety by the evanescently linking and unlinking foglets dispersed in the air. There isn't a mirror, either, although the fog simulating its reflective properties has done its job quite adequately, returning the appropriate light-rays to your eye. Or perhaps this isn't quite right; perhaps the only light you see comes from the foglets hovering directly in front of your eyes, and the image getting built up there from molecular pixels is just a kind of virtual reality fake. It gets hard to tell in Fog City and, really, it might not matter that much. Especially since your own body might now be built out of a shuttling coalition of exactly these clever foglets . . .

Are you still choking?

How can (physical) people breathe when the air is a solid mass of machines? Actually, it isn't really solid: the Foglets only occupy about ten percent of the actual volume of the air (they need lots of "elbow room" to move around easily). There's plenty of air left to breathe. As far as physically breathing it, we set up a pressure sensitive boundary which translates air motions on one side to Fog motions on the other.

Well, that's a relief.

Life is just a dream

Because so much can be simulated in an economy suffused by Utility Fog, physical reality will take on a dreamlike impermanence, a whimsical air a little like the toon-thronged fantasy world of the movie *Who Killed Roger Rabbit*. Toons, the physical manifestation of tools and agents and demons and objects (in computer parlance), will cavort and morph or simply vanish here and appear instantly over there. "The Fog acts as a continuous bridge," JoSH notes, "between actual physical reality and virtual reality."

Some caveats: you can simulate a saw buzzing through *faux* mahogany, but it won't work with the real thing. It can't deal with great heat. "A Fog fire blazing merrily away on Fog logs in a fireplace would feel warm on the skin a few feet away; it would feel the same to a hand inserted into the 'flame.' " Most crucially, foglets are *not* self-replicators.

We'll still require hard machines to carry us long distances—you can't really have a foglet composite carry you at the speed of sound across the Pacific—and to send big chunks of information. But a universal Utility Fog could deal with most of the daily needs of *Homo sapiens* on endless retirement. Every foglet is the equal of a 1990s supercomputer, as Drexler has shown in his analysis of nano computation, and "there are about 16 million Foglets to a cubic inch." Perhaps in their spare time, while they are simulating an empty room, these number-crunchers can be networked into a global effort to solve the Final Theory of Everything, forecast the remaining chaotic fluxes of the weather, cure cancer and death if those haven't already been dealt with, and write new episodes of *The Simpsons*.

But who controls the Fog? How do you stop the unchecked madness of living in a constant dreamy flux from unleashing the Monster from the Id? JoSH suggests some sensible protocols: "The Foglets within some distance of each person would be under that person's exclusive control; personal spaces could not merge except by mutual consent. This single protocol could prevent most crimes of violence in our hypothetical Fog City." Control of the Fog has to be massively distributed and decentralized, and robust where local pockets do fail.

Sturdier "industrial Fog" might be used for factories, fog composed of larger nanobots. If domestic fog has a strength and density somewhere between balsa wood and cork, "industrial Fog could have bulk properties resembling hardwood or aluminum. A nanotechnology-age factory would probably consist of a mass of Fog with special-purpose reactors embedded in it, where high-energy chemical transformations could take place. All the physical manipulation, transport, assembly, and so forth would be done by the Fog."

Foglets are powered by electricity, storing hydrogen as an energy buffer. And they wouldn't be terrifically costly to run. "If the Empire State Building were being floated around on a column of Fog," JoSH notes with a straight face, "the Fog would dissipate less than a watt per cubic centimeter." You might have some trouble getting rid of the waste heat, alas.[25]

Can we do it?

Given such delirious scenarios, backed by a decent amount of hard data-crunching, it might not be implausible, after all, that a medium-term minting technology, well short of the Spike, might allow people to drill into the Pacific or the Atlantic for their own private water supply. Optmistic Pelagius was still out there, arguing this case with pessimistic Augustine.

"Why would a mite-sized Exxon take orders from me, when the actual Exxon wouldn't?" asked crotchety Augustine, and Pelagius gave him back a soft but determined answer, one guaranteed to lift the hair on your neck as it sinks in: "Would you or I have any control over all this? I don't know. I never said we would be rich, or that we would even be around. I only said that the price of things will drop to almost, but not quite, zero. I am not at all convinced that we will survive the turmoil of the singularity, but we might, at least it's a chance, and without it we are all doomed to old age and oblivion. That is reason for hope," said Pelagius, a boundless optimist, "and at the very least, things won't be dull."

Perhaps the key to this debate is the *stickiness* and *balkiness* of the world. We inhabit a universe constrained by inertia. Critics with the transhumanist movement have doubted that anything like Vinge's runaway scenarios can occur, for this very reason. Professor Robin Hanson notes that superintelligent machines, the key to any nano Genie, might not arrive quickly via self-bootstrapping from computers that emulate human attainment. "I see no sharp horizon, blocking all view of things beyond. Instead, our vision just fades into a fog of possibilities, as usual."

And Dr. Max More, originator of the Extropian Principles, is equally judicious:

> The whole mathematical notion of a Singularity fits poorly with the workings of the physical world of people, institutions, and economies. My own expectation is that superintelligences will be integrated into a broader economic and social system. Even if superintelligence appears discontinuously, the effects on the world will be continuous . . . the speed and viscosity of the rest of the world will limit physical and organizational changes. Unless full-blown nanotechnology and robotics appear before the superintelligence, physical changes will take time.[26]

My own opinion? I think we are truly at the foot of a slope leading us into an entirely different society, one based on the extraordinary technological possibilities and freedoms and hazards we already see emerging from the labs, their curves converging as they rise. It is like sighting a hurricane on the horizon, moving toward us. We know it's on its way, even if the details are shrouded in darkness. Unlike a hurricane, the prospects of minting and AI and biological command need not be destructive. It's up to us to prepare for changes so drastic, coming at us so thick and so fast, that we can't yet truly imagine the shape of things to come. And that is the very definition of the Spike.

5: Climbing the Slope

THEY WERE FAMOUS PICTURES: *DEATH ON A BICYCLE, DEATH VISITS THE AMUSEMENT PARK.* . . . THEY'D BEEN A FAD IN THE 2050S, AT THE TIME OF THE LONGEVITY BREAKTHROUGH, WHEN PEOPLE REALIZED THAT BUT FOR ACCIDENTS AND VIOLENCE, THEY COULD LIVE FOREVER. DEATH WAS SUDDENLY A PLEASANT OLD MAN, FREED FROM HIS LONG-TIME BURDEN. HE ROLLED AWKWARDLY ALONG ON HIS FIRST BICYCLE RIDE, HIS SCYTHE STICKING UP LIKE A FLAG. CHILDREN RAN BESIDE HIM, SMILING AND LAUGHING.

—VERNOR VINGE, *MAROONED IN REALTIME, 1986*

Traditional Christian theology taught believers that the physical world would end in the Parousia, the return in glory of a resurrected Redeemer who would establish his new, clarified kingdom, reviving the dead so they might join those who survived the cataclysmic End Time battles, punishing the wicked in a Last Judgment. Modern fundamentalists still expect to see apocalypse explode outside their window any day now, despite its failure to occur on 1 January 2000, as their predecessors have expected it any day for the last couple of thousand years. They call it "the Rapture." Devotional art shows whole families rising into heaven inside their gleaming automobiles, into the arms of a smiling and radiant Californian Jesus.

The Spike is not the Parousia. Some of the dead, however, as it happens, might well be revived, if they've had the good sense to get their heads frozen. The living, too, might have a chance to enter a remarkable new kingdom, but it won't be the Rapture. Nobody expects Jesus Christ to be there.

You, on the other hand, might be. So might I. It's not clear what sort of shape we'll be in.

Vernor Vinge noted before the end of the millennium, in the

afterword to his novel set in a world emptied of all but a few stragglers following a Spike, *Marooned in Realtime:* "It's an ironic accident of the calendar that all this religious interest in transcendental events should be mixed with the objective evidence that we're falling into a technological singularity" (p. 67).

It's also a piercingly important distinction. Despite its reverberations with mythology and apocalyptic imagery, taken literally, *the Spike is not a religious vision.* It is a metaphor drawn from science, from the screaming upward curve of an exponential graph. That's all. Everything else is guesswork and projection, mixed, inevitably, with desperate hope and deep, archaic dreads.

On the other hand, as we've seen, the guesswork is not without foundations in reality. One of the paths into the singularity will probably be the explosive growth and power of minting technology, and that has already begun—if only in the smallest way, with those atoms shuttling their rudimentary calculations on a molecular abacus, with the CAD analyses of diamondoid gears and struts.

Putting a date on it

If there's to be a Spike—whatever that really comprises—when is it due? In his novel, Vinge deliberately set up a devastating plague war at the turn of the millennium to *postpone* progress, and still his Singularity came by shortly after 2210. "I showed artificial intelligence and intelligence amplification proceeding at what I suspect is a snail's pace. Sorry. I needed civilization to last long enough to hang a plot on it." In reality, as he made clear in his 1993 NASA address, he expects it by perhaps 2025 or 2030.

That's certainly the contrary of what most people will assume, learning that a Spike might peak in much less than a century. Impossibly rapid. It can't be true. New Age folly, they'll say. Wishful thinking of the dumbest, most self-serving kind. Vinge did not agree with that kind of dismissal in 1986 and, as we've seen, he has not changed his mind since. Instead, in the afterword to

Marooned in Realtime, he offered a prediction, meant as science fact: "If we don't have that general war, then it's *you* . . . who will understand the Singularity in the only possible way—by living through it" (p. 270).

The impact of artificial intelligence

While Drexler's portrait of a renovated world stresses the impact of nano minting, Vinge, as a computer scientist, has emphasized multiple breakthroughs in artificial and machine intelligence. Bringing those fields to fruition might very well require nanotech, needless to say, or at any rate comprehensive command of biology from the DNA level up (a feature of cutting-edge research that Vinge acknowledges). Either way, evolution will have passed out of the clumsy bumbling of accidental nature and into the purposeful domain of intelligence and imagination. We humans can *simulate* the effects of chosen changes—this, after all, is the function of foresight and planning and experiment—and thus bootstrap our designs thousands of times faster than natural selection can manage.

"From the human point of view this change will be a throwing away of all the previous rules, perhaps in the blink of an eye," Vinge declares, "an exponential runaway beyond any hope of control. Developments that before were thought might only happen in 'a million years' (if ever) will likely happen in the next century."

Why—to pose the question once more—must this feature of runaway change represent a singularity? Not just because it's a spike on the graph of technological progress, but owing to its transforming impact upon human reality in its entirety. The strangest feature of such a graph, taken literally—and Vinge does look at it with the straightest of faces—is that the higher you rise on its curve, the faster it climbs ahead of you. We can't catch up. We can't even get to the top and then slide despairingly back to the base. "As we move closer to this point, it will loom vaster and vaster over human affairs till the notion becomes a commonplace," Vinge points out. "Yet when it finally happens it may still be a great surprise and a greater unknown."

Dan Clemmensen says that it is strictly illogical to try to predict the date of the Spike as much as a decade in advance, let alone what life will be like (if there *is* life) after a singularity:

> (1) The singularity will be precipitated by the emergence of an internet-based, self-augmenting superintelligence (SI, hence SIngularity), and (2) This will occur within ten years of when I wrote "Paths to the Singularity" (i.e., before 1 May, 2006.)
>
> The most interesting new "insight" I've had is that it is illogical for anybody to agree with me, or with anybody who tries to predict the date of the singularity a decade in advance, whenever it may occur. The reason: technology is advancing "exponentially" or faster. This means that the bulk of the change in knowledge and capacity needed to precipitate the singularity will occur within the last year before the event. [Personal communication, September 1996.]

In that sense, which I find persuasive even if the dating was preposterous, our enterprise in this book is both quixotic and impossible. It is—to return to the inevitable religious comparisons I'm trying so hard to skirt—akin to the futility of a theologian or a physicist attempting to understand the Mind of God (as Stephen Hawking rhetorically dubbed his own scientific efforts).

The emergence of ultraintelligence

The great mathematician John von Neumann, who almost single-handedly laid down the foundations for modern computing before the Second World War (Alan Turing is the other candidate for parent of both the computer and AI), seems to have glimpsed something of the same heart-stopping prospect. His friend Stanislaw Ulam quotes him thus:

One conversation centered on the ever accelerating progress of technology and changes in the mode of human life, which gives the appearance of approaching some essential singularity in the history of the race beyond which human affairs, as we know them, could not continue.[1]

Although von Neumann invented the computer we know today, he failed to see that beyond conventional slopes of progress was a dizzying rise into superhuman intelligence, biological or contrived in a machine. It is this jump to ultraintelligence, especially through computer implementation, that leads into a kind of explosion of smartness. Would such a machine be "docile enough to tell us how to keep it under control," as British mathematician I. J. Good suggested in 1965?[2] Most unlikely.

During the immediate future, we'll see the hardware doublings discussed previously. Machines will get faster and smaller. Both their random access memory and file storage will expand prodigiously. At some stage, the raw processing power of some machines will equal that of the human brain. Quite a short while later, this will extend to smaller computers, even notepads.

No mind, no gain

Yet a mechanical brain without a mechanical mind is just a very fast abacus. A mind, in these terms, is a *process*, the outcome of suites of specialized software that manipulate and integrate information from the outside world, feeding back the results, eventually, into that external reality.

It's plausible that writing the software for such minds will be far from straightforward, Vinge notes, "involving lots of false starts and experimentation. If so, then the arrival of self-aware machines will not happen till after the development of hardware that is substantially more powerful than humans' natural equipment." Sheer processing power will have to take the place of those subtle algorithms, many of them built in to the cellular architecture of the brain, that evolution has sorted and retained during literally billions of years of shuffling and testing to death.

But this implies something striking, almost unprecedented. A brain potentially smarter than the mind it is running! Once machine minds begin redesigning *their own software,* we can expect to see shockingly fast developments. Right from the outset (it's already happening), computer-assisted design programs will take part, of course, in coding the tedious details of systems architecture, but I mean something more profound. The precedent is the human brain itself. Maybe it was the consolidation of specialized grammatic structures in the cortex, perhaps it was a certain reentrant looping that let some parts of the brain listen in on other parts—whatever happened, brains emerged on earth that were able to self-bootstrap through culture, using shared learning and trial-and-error and systems of ideas held not just inside a single skull but dispersed among dozens, then hundreds, then tens of thousands . . .

How much raw information and processing power would it take to emulate a brain, to forge an artificial or machine mentality?

Drawing on current neurophysiology, Frank Tipler recently estimated it this way: we possess about 10 billion (10^{10}) neurons or brain cells in our heads, each of them with perhaps a hundred thousand (10^5) connections at their synapses to other neurons all through the brain. If every synaptic link encodes a single bit of information, that's 10^{15} fundamental units of knowledge expressed in on-offs, the binary language of computers. Since we're trying so hard to be conservative (really, I mean it), make that 10^{17}—a hundred thousand trillion bits.

Now, brain neurons tend to be active between one percent and ten percent of the time, firing about 100 times a second. If one activation equals a single act of cerebral calculation, and a single neuron requires 10 million individual computer calculations to simulate it, then at worst a living brain can surely be emulated by a hundred thousand teraflops (or trillion computer calculations a second).

Can we do this? Not yet. Not nearly. As we saw earlier, machines already exist that do several trillion calculations per second. So our current devices are 33,000 times too feeble for the job. We need to increase computing power from 3 teraflops to 100,000 teraflops (or, more concisely, 100 petaflops)—a whopping factor. Impossible, surely?

Not at all. Exponential growth cuts numbers like that down to size. Exponential growth, the secret of the Spike! A mere fifteen doublings meets that goal nicely. Here's the sequence that multiplies your starting point to about 33,000 times the original figure:

In the first period, you go from 3 teraflops to 6. In the next you go to 12 teraflops. Then to 24, then 48, 96, 192, 384, 768, 1536, 3072, 6144, 12,288, 24,576, 49,152, and finally, at the fifteenth doubling, 98,304 teraflops, close enough to the 100,000 we think we might need at the outside.

With computer power doubling somewhere every 18 months or faster, we can expect this in . . . oh, 25 years, tops. What's more, Tipler noted, suppose his estimate of the upper bound needed to create a human-equivalent brain is wrong, too low by a factor of a hundred. Why, then, with this fabulous doubling going forward, it will *still* only take an additional decade or less to develop machines with enough speed and memory to emulate a human brain. So to put a date on it: somewhere around 2025 or 2035. The usual suspects.

Building a mind

Does that mean we will also be able to emulate a conscious mind, as well as a human-level number-crunching brain? Can we even pose that distinction any longer (assuming the necessary basic algorithms, or "grammar," are in place)? Is a developing mind something more wondrous and mysterious than a working, well-stocked brain? A brain, that is, that is coupled satisfactorily by sensory channels to the external world, and with appropriate effectors (voice and hands and feet, say, or their machine counterparts) to influence that world?

Schooling is a kind of self-implemented software redesign. So is the work environment, once it breaks free of sheer rote repetitive drudgery. So, too, are song, writing, reading, mathematical tricks of coding the world into abstractions easier to handle. Now we humans have come pretty close to the optimum use of our current brains, or at least the finest thinkers and artists have done so, and we are stalled.

Machine intelligence, with its self-organizing minds run on better-than-human hardware, will burst through the bottleneck. And there'll be no stopping that growth, because unlike human minds these new creatures will not be limited by the constraints of conservative DNA, that fabulous molecule that holds patterns almost perfectly preserved down the river of time. That is why evolution takes such a *long*, agonizing time to ratchet itself up a notch. Break free from that mindless process, via machine self-programming or genetic engineering or both, and everything changes *fast*.

How fast? Vinge is uncompromising: "The precipitating event will likely be unexpected—perhaps even to the researchers involved. ('But all our previous models were catatonic! We were just tweaking some parameters. . . .') If networking is widespread enough (into ubiquitous embedded systems), it may seem as if our artefacts as a whole had suddenly wakened."

After that, after the first machines have awoken? Nobody knows. This note of caution is worth attending to, since it comes from an unexpected source: "For all my rampant technological optimism," Vinge himself states, "sometimes I think I'd be more comfortable if I were regarding these transcendental events from one thousand years remove . . . instead of twenty."

Sleeping through the Spike

In his influential novel *Marooned in Realtime,* Vinge placed his characters at a fifty-million-year remove from the Spike—in the remote future, after the Singularity had passed across the world, leaving only traces of vast, unintelligible engineering works and decaying cities. How these inadvertent survivors managed their delayed second chance, and what became of them, makes an intriguing tale. What's truly arresting, however, is the dark place at the center of that imagined history. You wake up to find that the Parousia has come and gone, Gabriel's Trump has blasted its brazen song and you missed it . . .

I have been insisting that the Spike is *not* any kind of Second Coming, but that's not to say that it won't be a sort of immanent transcendence, an accelerating dash into incomprehensible glory. (Or ugliness, always a possibility: nano gray goo, viral green goo, a thousand varieties of unbalanced superminds let loose in the playground of the solar system . . .) Here and there in Vinge's novel are hints of what might have occurred when humanity vanished clean away while the reader's back was turned. " 'Mankind simply graduated, and you and I and the rest missed graduation night,' " one character tells another. " 'Just talking about superhuman intelligence gets us into something like religion. . . .' "

Still, how fast could these millennial changes really happen? Even in a swiftly changing world, another character muses doubtfully, "there had been limits on how fast the marketplace could absorb new developments . . . what about the installed base of older equipment? What about compatibility with devices not yet upgraded? How could the world of real products be turned inside out in such a short time?" Vinge seems here to start interrogating his own cool idea only to back hurriedly away. It is a piece of narrative flim-flam in the guise of rhetorical questioning.

We, however, need to pause and take stock. Is geometric, or exponential, growth all it's cracked up to be? Prophets of doom are eager to invoke its steeply rising curves, linking them to dire warnings of imminent resource depletion, human overpopulation, every manner of self-inflicted woe. Some of these hazards are not easily dismissed, although the Spike will probably make most of the warnings laughable in retrospect. Still, the ready way in which such growth patterns are turned into crushing juggernauts needs to be investigated with at least a grain of skepticism.

You can't keep growing forever

When scientists tell you that a bacterial culture placed in a test tube full of nutrients will double and redouble every fifteen minutes until it fills the available space within mere hours, this is doubtless true. Microorganisms have a frightening way of chomping their way through all the accessible food sources, turning every scrap of free nutrient into more germs. Rhetoric jumps from this simple truth into partial truth, and thence to nightmare and finally absurdity, all without blushing.

Imagine a world ocean of nutrient soup, such as might have formed early in our planet's history. Introduce a single replicating molecule, either by spontaneous self-assembly or dropped in (as Sir Fred Hoyle insists) from space. The thing will double and redouble and fill the earth, literally. Eventually, of course, it will choke itself, and a new game will open up, in which the only way forward is through one critter cropping another: the birth of the food chain.

Rhetoric, however, typically ignores this fact of limits to growth. The mathematical curve is abstract, and knows nothing of the earth's gravity, its spherical closure, the difficulty of escaping from its atmosphere into the larger universe. So when boisterous analogies insist that unchecked geometric growth will result, after a few years or decades, in a ball of replicating bacteria expanding at the speed of light, you have to stop and say: don't be ridiculous. The metaphor is what has escaped from planet Earth, not the multiplying microorganisms.

Yet the analogy has proved terrifically seductive to propagandists of every stripe. Consider the arguments of Dr. Isaac Asimov, who certainly knew numbers and biology and physics. In the example that follows, Asimov is *not* claiming that human populations will surge toward a literal Singularity. Indeed, it's that *reductio ad absurdum* that his true argument hangs upon. He wishes to convince us that there are indeed limits to human growth, and that the sooner we do something about it the happier we shall be in the long run. Here is his analysis:

The world's human population is currently doubling about

every 35 years, because on average it is growing at two percent each year. By 2049, we might expect there to be some 16 billion people alive, many of them in parlous conditions (Asimov does not make allowance for the miraculous benefits of nanotechnology). While he does consider the impact of technological improvements, he notes: "It is easy to show with absolute certainty that the present rate of population growth, *if it continues,* will easily outpace not only any likely technological advance, but any conceivable technological advance."[3]

This is an extraordinary claim. Outpace *any* technological fix? Yes indeed, for exponential factors assure this result. Suppose by 2100 humans number 50 billion. All other animal life is gone, replaced by people. No plants exist that are not entirely edible. Such a world, Asimov calculates, might support a maximum population of 1.2 *trillion,* a peak reached just 300 years from now. But let us follow the curve. Doubling onward every 35 years, within less than two millennia "the total mass of humanity would equal the total mass of the Earth."

This seems self-evidently absurd, even if we allow that minting might take the world apart all the way down to the core and recast its atoms into human flesh. Indeed, a large part of that vast population could happily exist in bubble habitats surrounding the sun in a vast ring, filling all the orbit currently swept out once a year by our planet. But let's not stop there. The total mass of the known universe, Asimov estimates, is greater than the Earth's by a factor of . . . 5,000,000,000,000,000,000,000,000,000. That is, Earth times 5 by 10^{27}. Could we chew up all that mass and turn it into people? Certainly, if the exponential curve is taken literally. Doubling each thirty-five years, it would take— what? A billion years? A hundred million? Inconceivable aeons, surely.

No. Think Spike. It would take just five thousand years, in principle, to eat the universe.

Expanding faster than light

What's wrong with this argument? Well, it contains a fundamental error. Even with fabulous nanotechnology and machine intelligence, this preposterous scenario assumes that people can disperse across the universe, reproducing as they go, at far greater than the speed of light. Well, sadly, to the best of our finest knowledge, that is simply impossible. Some loopholes in relativity's equations have been discerned and one can suggest without complete idiocy that "wormholes" might some day be created or found that permit travel (or signaling, at least) at superluminal velocities.

Still, it is quite ridiculous to project a volume of humans roaring outward faster than light, consuming everything in their path, and peaking in a kind of final, lethal population Spike in A.D. 7000. That is just a plain mistake of analysis, taking the graph for the reality. (Aside from anything else, this scenario seems to prove that no other species with a population doubling-rate remotely similar to ours can exist anywhere else in the entire cosmos—or they would already have turned the solar system, and us, into themselves.)

The wild, wild future

The question is: does the Spike postulate make such an error with regard to the much more limited runaway growth patterns we discuss in this book?

A bacterial culture can't expand at unlimited speed because it can only eat so much, so fast, and it can spread rapidly only at the edges of its petri dish. Even with laser-propelled starships and terraformed asteroids of the kind promoted by technological optimists, the human population will keep hitting saturation points that balk its spread for long periods. Why shouldn't the same be true of the soaring curves that, taken naïvely, appear to promise ever-smaller, ever-more powerful computers and nano fabricators?

Yet some of the steps that make this headlong alteration thinkable, if not altogether feasible, are sketched in Vinge's *Marooned in Realtime.* "High-tech" people from close to the Singularity wear headbands that augment their native abilities, computer patches added to the raw stuff of evolved brains with their limited memories and narrow attention windows. (More up-to-date scenarios might expect such chips to be surgically implanted deep in the brain, or perhaps to be grown there using engineered cells or nano constructors. In a way, it's evidence of Vinge's own case that a mere fifteen years later we already find many of his once-wild projections rather tame and unadventurous, in the light of today's cutting-edge science and technology.)

Why would we need such amplifiers? Well, for starters, even the smartest of us can hardly maintain a conceptual grasp on more than seven different items at once, and five is the usual mental handful. So we live in a cognitive universe dimly glimpsed through the narrowest of cracks, and the width of that aperture is set by our inherited neural hardware. Boost it, link our augmented minds together, and who knows what wonders of awareness might burst open into consciousness?

"Humankind and its machines became something better," speculates the most advanced of Vinge's high-techs, "something ... unknowable." Yet his own experience, at the opening of the twenty-third century, is almost incomprehensible to us at the turn of the millennium. With his seven colleagues, he was engaged in mining the sun for antimatter, "distilling one hundred thousand tons of matter and antimatter every second. That was enough to dim the sun, though we arranged things so the effect wasn't perceptible from the ecliptic"—the orbital region around the sun that contains the earth and other planets and most asteroids. Working so far from home, he and his companions were brutally severed from the real action, "hundreds of light-seconds away."

You may smile at this wry jest. Pause, though, to reflect on what such a world might be like if it truly existed. Two hundred years ago, Europeans explored and conquered large parts of a single world many months distant from their political and mercantile masters at home. Today, by contrast, nobody with access to a phone is more than fractions of a minute from anywhere else on the globe. Computer networks, swapping information and financial transactions, blur into a haze of virtual instantaneity. Imagine a world where people living halfway across the solar system are unalterably amputated from the current action, because the speed of light is an intractable barrier to faster communication.

On Earth, in Vinge's future world, large corporations with better computers merged their staff into linkages of thousands. Is this a horrible prospect of soul death, extermination of the self? One might expect such an interpretation from a libertarian like Vinge, but in fact he suggests otherwise: "There was power and knowledge and joy in those companies..." As the Singularity approaches, the mind-to-mind linkages and augmentations become extreme, forming a group mind: "By the beginning of the twenty-third, there were three *billion* people in the Earth/Luna volume. Three billion people and corresponding processing power—all less than three light-seconds apart."

It might seem foolish to spend so much time on what is plainly a contrived, if enjoyable, piece of wild fiction. After all, I am claiming that the Spike is more than a literary conceit, much more than an idea that's escaped from science fiction. Indeed that is so, but it would be vulgar and narrow to ignore Vinge's speculations simply because he has chosen this entertaining way to explore them. Think-tank futurists use such scenarios precisely because they engage our imagination in depth. Stories are not automatically contemptible because they are invented. They are, and always have been, the way we think about possibilities, the way we try to grasp the consequences of what-is-not-yet or, more generally, what is *not yet understood*.

Climbing the steps of intelligence

People often say that an enhanced intelligence will relate to ours as we do to animals, even bugs. But in important ways human mental life is not really different in kind from the progressively

more limited sorts found in dogs and cats and parrots and gold-
fish and slugs and sunflowers and bacteria.

Of course, many people argue that our minds *are* exactly dif-
ferentiated from those of "lesser creatures" in some mystical, un-
approachable way. I am not persuaded that this is a fruitful
prejudice. Even supposing that it turns out to be true—that people have an extra gadget . . . let's call it a *soul* . . . that provides our
self-awareness and radical creativity—that is a conclusion to be
attained only after long, scrupulous investigation. It would be, in
a sense, a negative hypothesis: something we'd be forced to en-
tertain purely because of the failure of more rational models that
are compatible with everything else we know about the physical
universe. (And trust me—we now know a *colossal* amount, and
most of it is tied together into patterns not easily edited or expanded without shredding the lot.)

So let's assume, because this seems the most provident starting
point, that human minds are not *mysterious things* but are rather
very complicated processes going on inside living brains connected
by senses to the outside world. Is it reasonable to guess that
doubling the power of a laptop computer, and doubling it again,
and redoubling repeatedly for another fifteen or thirty years, will
automatically produce first a smart machine, then an intelligent
machine, and finally . . . a *hyperintelligent machine*?

Vinge poses this rather neatly: "Imagine running a dog mind
at very high speed. Would a thousand years of doggy living add
up to any human insight? (Now if the dog mind were cleverly
rewired and *then* run at high speed, we might see something
different . . .)" No computer is currently anywhere near the
doggy level of cleverness, of course, but some of them approach
the simplest insects. Might running an insect brain very, very fast
turn it into someone you'd like to discuss politics or art with?

It seems obvious that the answer is no, but what if that insect
brain joins a swarm of its fellows, and they learn (or are taught)
the trick of specialization, so that one bunch of insect brains
swaps information about how stuff *looks* today, from a number
of strategic angles, and a second bunch focuses on *sound vibrations,* while other bunches of ant brains store these impressions

and disgorge them on command from yet other groups . . . I have just propounded the principle of the hive, which we know exists. And in many ways the hive is much cannier than any of the dumb modules with legs and wings that comprise it.

You might see that I've also just sketched something like a complex brain—like a *human* brain, in fact. Each of our neurons is a little like an insect (simpler, in fact, and much smaller), but it has better lines of communication than any single ant. Take a hundred thousand million of *those* guys and let them cross-wire to each other in some kind of self-organizing hierarchy, and you do indeed get . . . *us.*

Speeding up a dog's brain

Even if you added in a big pack of extra memory, a dog's brain lacks the architecture to think like a person. Well, what about an immortal and massively augmented dog, genetically engineered to have as many cortical neurons as a human, say, but arrayed as they'd otherwise be in a normal dog. I suppose it's remotely possible that this pooch might, perhaps with great inner agony, teach itself the tricks of self-consciousness. It would need access to a brainy culture, clearly enough, and right now we are the only one available. So this hyperdog would have to have eyes capable of reading, and paws dexterous enough to handle pages or keyboard, and a jaw and larynx rewired for speech. It's a big job. Almost certainly no such wonder could emerge from a single, extreme mutation that just pumped up the puppy's brain capacity in utero.

A computer, though—That's a different story, since it's much easier to reallocate uses to machine memory, to rewrite code and try it out and discard it when it bombs, and then try again, and keep trying until you get it right. That's presumably true even if the machine is bootstrapping its own abilities, driven away from stolid stable inner states by random changes somewhere in its operating system (in the software, that is, which makes it a process rather than just a lump of expensive silicon). This is the evolutionary model of teaching a machine to be a person. In a sense, we know that this has to work, because on a geological timescale it is what produced us. Mutation, contest, and cooperation, differential survival of genetic patterns according to the success of the bodies they built: natural selection, in a word.

Amplifying human intelligence

We have been looking at the path to AI, or artificial intelligence. There is a quite different method of attaining advanced cognitive abilities: a switch in emphasis that is captured deftly by switching the name to IA, or *intelligence amplification*. Now it's the human brain that is being boosted, or linked to others of its kind in a group mind. Again, this is eerily familiar, precisely because it is a definition of a society, or a culture.

Already our brains store much of their knowledge outside the skull, in books and film and magnetic tape, in the huge hard structures of cities and aqueducts and farms, in the facts and opinions we can get by talking to each other. Vinge suggests that the path from this existing state of affairs to enhanced intelligence will be rather easier than the AI route, because we have actually done it a number of times without quite realizing it. Currently, access to the Internet through a fast workstation is a genuine augmentation of a researcher's capacity: "in network and interface research there is something as profound (and potential wild) as Artificial Intelligence."

In effect you can now think faster, put information together more swiftly, send your results out without delay, gain the reciprocal benefit from others cruising the network. Once improved methods are devised for getting information out of a database, and shaping knowledge without the need for writing or keyboarding or even talking—I mean something like the cyberpunk dream of "jacking in to cyberspace"—well, it will be a whole new kind of life. One small step for the individual cyber-surfer, one mighty swarming leap toward the Spike.

So much for metaphor. What of real technological attempts to attain these unprecedented feats of self-mutation? Can we boost our minds without losing our souls, and our world?

The Internet wakes up

One method that might already have begun its insidious Trojan Horse progress is the merging of the global Internet as a mind/machine interface, a development Vinge believes "is proceeding the fastest and may run us into the Singularity before anything else." An extreme version has been suggested by systems expert Dan Clemmensen, who expects a single luckily placed researcher or hacker to bring the Net to life, as it were.

Suppose the increasing numbers of users of the Internet fetch its linkages to some "critical mass" of interconnectivity, so that it . . . *wakes up*. Raw computing power is not all that's required, Clemmensen notes. "It's possible that the final missing link will be a particular piece of software such as an information-visualization package, or a decision-support package, or a knowledge database. The point is that the Internet may then enter a super-critical condition in which a single seed may precipitate a phase change"—as very cold water can crystallize into a new form as ice. The seed program would borrow (or steal) computing resources to augment its own intelligence.

If this seems ridiculously optimistic, consider Clemmensen's "SuperIntelligence Dream" scenario. The doubtful might well regard this as a nightmare, assuming it's not altogether preposterous:

> a researcher (probably a grad student at MIT, drinking Jolt cola at 2 a.m. and programming when he should be studying for an English exam) is attempting to enhance a decision-support system by interfacing it to a knowledge base and to a graphical information-presentation system. Because he's interested in software development, the knowledge base is the one he set up last year as a class assignment in his software engineering class. He gets the system up, and

(since he is currently working on this system) his first trial run is an attempt to optimize his prototype. He succeeds, and installs the next version. With this version, he optimizes the operating system.[4] Next, he optimizes his hacking program. He grabs all the work-stations in the dorm, via the net, and optimizes them. Then he reoptimizes his program to run in a distributed mode. Now (about 4 A.M., I think,) he hacks the campus routers, and then all computers on the campus, and then the web. He turns his attention to extending his knowledge base, probably by hacking the CYC database. By 6 A.M., he's running in the whole Web. By the end of the trading day, he owns a controlling interest in a nice collection of companies on the New York Stock Exchange.

What we have here, you'll have noticed, seems less like a super-intelligent machine, or even human/AI hybrid, than a perfectly ordinary person with a powerful tool at his disposal. Clemmensen goes further, though, calling the combination "a human/computer collaboration whose intelligence is substantially augmented by its computer component. 'Intelligence' for the purposes of this discussion is very (!) narrowly defined as the quality that permits an entity to design and implement newer and better computer hardware and software."

What's more, he notes that if an entity with superior intelligence can augment its own abilities faster than a stupider entity (one as stupid as you or me, say), this constitutes a fast feedback loop. Here's the recipe: human, plus the tremendous distributed hardware underlying the World Wide Web, plus novel software composed of filters and agents able to sort and combine huge amounts of data, plus knowledge bases such as CYC (an existing and growing "natural language" encyclopedia constructed by Douglas B. Lenat and many helpers). If Clemmensen is right, this blend will be more than human. What's more, it will have the capacity to bootstrap its own intelligence to higher levels, faster and faster as it gets smarter and smarter.

Others are skeptical of this scenario. Economist and political scientist Robin Hanson notes that individual computers have been linking in to the Net for years now, and there hasn't been any conspicuous takeoff. He regards the suggestion as "wishful thinking." Many will agree with him. Some deny that the scenario is even possible, because the hundreds of thousands of individual systems comprising the Internet are well protected from invasion, defended behind what are called "firewalls." It doesn't mean, however, that some day—perhaps tomorrow—the critical level of connections will not be achieved, the right mix of human smarts and knowledge base and program package come together to gel into a self-bootstrapping superintelligence. What then? Vinge has observed: "Even the egalitarian view of an Internet that wakes up along with all mankind can be viewed as a nightmare."

Fear of intelligence

Must the Spike, seen as the amplification of human intelligence, or its replacement by machines, be a horror story? No doubt this is the way it will be portrayed by Hollywood. *The Blob, 2001. I Married a Supernerd. The Terminalator.* Whenever this prospect has been broached by the mass media, the usual phony contrasts have been thrust upon us. Mind versus Passion. Love throttled by Rationality.

This is a very strange cliché. Its supposed truth is denied by the slightest contact with real, frenzied, passionate human scientists and technologists. But perhaps this is what *really* frightens us. What happens when strong feelings, devotion and hatred and prejudice, are joined powerfully with effective strategies for influencing the world? Sometimes we get holocausts, the risk of global nuclear war that could serve nobody's purposes, mad cults with real weapons.

Is the Spike likely to bring just such unholy conjunctions to a new pitch of fearful strength? We have evolved in hundreds of thousands, perhaps millions, of years of sluggish tribal life, dealing with each other face to face, vulnerable to the fists and scorn of others, sensitive to their smiles and touch. What happens when we allow ourselves to deal with others wholly through a monitor's window?

Anyone who has watched, or taken part in, a "flame war" on the Internet knows how these exchanges escalate into brutal rudeness and sarcasm, the kind of thing that leads on the real highway to "road rage" murders. It is a stroke of luck that we can't yet easily kill our foes through the screen. But perhaps an enhanced human, locked into symbiosis with potent programs running on computers all around the world, would indeed learn how to smite enemies in world-shaking tantrums.

With luck, this interregnum would be brief. By and large, people are not mad dogs. It is already possible, and has been for decades, to make lethal bombs from common agricultural substances, or gun down children in a playground, yet when this is done by a few crazies the rest of us are genuinely grief-stricken, if only for a brief time before our own concerns drift back to the surface.

The intelligence amplification path to the Spike

The transition from intelligence boosting to the Spike, when enormous numbers of changes happen with immense swiftness, will probably not occur as an all-at-once crystallization—or at least we must hope not. On the other hand, transition into superintelligence could bring with it the solutions to most of our traditional woes: hunger, thirst, nakedness to the elements—scarcity, in a word. The means to those solutions might not even be glimpsable yet, but the paths we have examined so far suggest the kinds of methods an augmented culture will use routinely: nanominting, direct AI-interfaces, genome control and repair, massively extended life span . . .

In such an authentic golden age, immortality would be no longer a fantasy of consolation for the imagined world beyond death, but a literally indefinite extension of life in a utopian world without want. But somehow that tale chills us, and with good reason.

We are aware of our limitations, the weaknesses of brains and bodies built by blind Darwinian selection. Each of us expires after a fertile span of some thirty or forty years. We are put together as disposable gene carriers. Our beautiful minds rot with our fallible, corruptible brains, and are gone forever. What would happen to such temporary mechanisms if they were repaired again and again, held safe from the corrosion of time? Nightmare, perhaps. Endless repetition. So, to be tolerable, *extended* life must also be *enhanced* life. For immortality we would need to be smarter, even if we didn't already need amplified intelligence just to *attain* immortality. (In fact, we might not need it, as medical consequences of the Human Genome Project could conceivably show the way to endless cellular repair.)

More to the point, what is the "we," the "I," that is going to survive into the post-Spike utopia? If fashionable theorists insist that the self is always already a construct, an illusion, this perspective will be even harder to dispute in a world where we can send out "partials" or "agents" from ourselves into the global Net. Already, our brains are composed of dedicated, somewhat partitioned modules resembling the "faculties" of an older philosophy. Of course these specialized components tend to work together, weaving (most of the time) a sense of unified consciousness. Once we learn to split off fractions of our selves—or, rather, to duplicate and amplify and elaborate those fractions—we will no longer be strictly human. Vinge, as always, is there ahead of us: "These are essential features of strong superhumanity and the Singularity. Thinking about them, one begins to feel how essentially strange and different the Post-Human era will be—*no matter how cleverly and benignly it is brought to be.*"

The grace to permit remnants of the old, unreconstructed humanity to live in peace may be the best we can hope for from augmented or group-mind posthumans. Well, maybe one happier option is conceivable, if you are open to the charms of self-deceptions: our successors might choose to disguise themselves as our servants. Vinge hints at "benign treatment (perhaps even giving the stay-behinds the appearance of being masters of god-like slaves)." Unaltered humans would be the metaphysical equivalents of the Amish, those serene agricultural throwbacks in our mechanized, electronic world. That appears to be a difficult way of life to sustain, even to negotiate, and its future version would surely be harder still to endure without soul-wrenching episodes of blatant bad faith, backsliding by the next generation, lack of resolution when times grew tough and one's body aged while everyone else remained healthy and young.

Many paths to the Spike

We have considered two great highways into the transhuman and posthuman futures of the Spike. The first is machine intelligence, even consciousness, of a type not merely equal to our own but superior to it. The second is augmentation of existing minds and bodies, so that flesh and metal and silicon bond into superhuman cyborg forms. Later we'll return to these possibilities and explore them in greater depth, but for now it is necessary at least to sketch out two rather different paths to the Spike.

In one possible world, machines simply self-bootstrap right past us, moving under the impulse of their own modifications and recodings into a level of sentience that flesh can never hope to attain. This makes sense, despite the affront to our pride, precisely because AI replaces slow neural structures—sluggish chemical neurotransmitters, ionic currents, and hardwired genetic design—with blazingly fast electronic or photonic neural nets that can be rewritten at will, and memories constrained by the size of our skulls with hard storage limited only by speed-of-light access paths. That likelihood has been allegorized in fascinating detail by the brilliant Polish generalist Stanislaw Lem, notably in a parable called "Golem XIV." A superintelligent oracle devised during the Cold War, Golem XIV is about to Transcend into "uncompromising silence."

A second possible world blends human and machine not by prosthesis (that is, enhancing our current bodies with chips, modules, and interface devices), but in the opposite direction. We could *leave* our bodies, *become* machines, physically transfer our minds into computer platforms. If earlier versions of the Spike remind us inescapably of the Christian myth of the Parousia, this one is eerily akin to religious hopes from Asia: reincarnation, where an impalpable essence slips out of an aged body to enter the waiting vessel of an unborn infant. Or perhaps of related beliefs, in which each spark of consciousness is a piece of the Godhead, the physical world mere *Maya* or illusion . . .

Yet today's transhumanist ambitions have nothing significantly in common with those ancient dogmas. Transhumanists are materialists who maintain that mind is indeed nothing other— though certainly nothing less—than the sublimely complex workings of the physical brain and its bodily extensions. If that is what we *are*, why, what is to prevent us from copying—mapping—our neurological complexity into some more durable, swifter material substrate? This process, called "uploading" by some and, confusingly, "downloading" by others, is a path to the Spike in which we *can* take it with us, but only by leaving everything behind. A daring insight worthy of a Zen master.

Some will recoil in horror. A brain in a vat, the cliché from bad movies. What adolescent silliness this all seems. How vile, yet how predictable, to see such bloodless proposals emerge from the ranks of the technophiles, the body-hating, frightened computer hackers, the social incompetents in retreat from sensuous reality. Well, yes, surely this charge has some merit to it. Several of the proponents of uploading make no bones about their distaste for the limitations and messy urgings of the body and its Darwinian drives. But the farther you go into this strange territory, the less you can explain away the idea as simple hatred of the flesh.

The mind-body problem

If uploading looks at first like the latest incarnation, so to speak, of classic Cartesian dualism, where mind was deemed to be finer and truer than body, why, then, so too does its opposite. Let us suppose that mind and passion and *soul* are indeed the body, that whirling composite of matter and force and energy, in motion in the world. How does it do dirt on any of these worthy phenomena to imagine mind changing its habitation from one kind of organized matter to another?

We are all too readily trapped by preconceptions from earlier decades, earlier *centuries* if it comes to that. Is the mind a machine? Certainly not, we shout indignantly, thinking of clocks and washing machine motors and even, perhaps, the stupid computers on our laps. But those limited personal computers are already a far cry from a wind-up parrot, or a piano driven by a paper tape. The kind of machine our mind is like is a Turing machine, a universal computer with vast memory, able (in principle) to mimic any set of logical steps and causal connections. It is not like a broken-down lawn mower, and nobody ever thought it was.

Likewise, when people misguidedly complain that enthusiasts for virtual reality are in antiseptic flight from the body, they stumble at the very first step. Yes, VR as currently implemented is rudimentary, grainy, coarsely colored, slow. Yes, the immersion we can achieve is restricted to an unconvincing visual display, perhaps in three insecurely rendered spatial dimensions, plus a minimal tactile dimension available through data gloves and reactive joy sticks. But that is a transitional limitation. VR won't remain like that for long.

To insist that it will is like someone at the start of the twentieth century listening grudgingly to a toy crystal radio set and denying the possibility of rich orchestral CD stereo. Virtual reality is intriguing, perhaps epoch-making, exactly *because* it links abstract computational data structures with our inherited senses, with our *bodies,* because it makes us feel and see and eventually smell and taste the denizens of cyberspace. VR, despite what you might think, looking at certain pimply, unwashed, pizza-gorging devotees, is not a flight away from the body—it is the finest medium yet devised for allowing imagination the theatrical texture and density of the flesh.

Staying in touch with reality

Uploading need not imply a world of pale, bloated grubs lying in the dark with their brains wired to spreadsheets and simulated worlds like *Myst*. Extropian philosopher Max More, who intends to upload when that becomes an option (and use his new freedom to explore the stars), puts his own case: "I'm in the gym five days a week, plus I either run or cycle. I can boast that I do 710 pounds on the leg press. No atrophied body here!"

The initial goal of uploaders would be to emulate the brain by superior means, and that requires connections to reality, it calls for give-and-take, it builds from the peculiar truth that inside our pinko-gray brain matter is where our selves are generated. That does not denigrate the body, far from it.

A paraplegic with no more access to the world than her mouth and ears and eyes and her vivid, courageous brain *is a person*. By contrast, the superb corpse of an Olympic athlete or concert pianist with a fatal brain injury, metabolism sustained by a heart-lung machine, is no kind of person at all, just a tragic reminder of the fallibility of life and a storehouse for luckier transplant patients.

Copying the brain

Suppose a neurosurgeon could isolate a single cell in your brain or some other part of the central nervous system. It's a queasy thought, but picture your naked brain exposed to the operating theater as this neuron is selected with extreme care, in no way damaged by the instruments examining its function. We might wish to imagine this being done by Drexlerian nanomachines, barely larger than viruses, swimming in and around the cell, recording every detail of the cell's architecture and workings.

Using this data, an adjacent crew of nanites might construct a tiny machine on the same scale as that neuron, able to accept the signals flowing in along its synapsing dendrites and out along its axon, processing the coded pulses just as the neuron itself would do, and emitting precisely the same kinds of messages that it would send the rest of the brain.

Carefully, wire the redundant surrogate into the brain's neural net. Switch signal tracks back and forth, calibrating the two. Finally, when you are quite sure that the artificial neuron is performing exactly as the nerve cell it's copying . . . snip out the original neuron and discard it.

Has anything crucial changed?

By definition, no, or you would not have felt confident in completing the replacement. We are not talking here about scissoring out the soul, or even a fragment of the soul. The small replacement is doing exactly the same job as its model.

Having convinced yourself of this fact, do the same thing another hundred billion times.

Close up the skull, return the patient to recovery, and wait anxiously for the moment of awakening.

The patient's original organic brain is altogether gone. Where neural tissue filled a skull, now it works by superbly contrived copies in silicon and other materials not known to evolution. Can you tell the difference when the patient wakes, looks up, recognizes you, speaks your name? Can the *patient* tell the difference from the inside? Even if neither of you notices anything out of the ordinary, is this still nothing better than a diabolical illusion, a shockingly slow murder disguised by the monstrous humanoid contraption left in its wake?

You must understand that this is nothing more than a crude thought experiment, almost a *reductio ad absurdum* that fails, rather surprisingly, to reduce to absurdity. True, replacement of neurons on a one-by-one basis would be preposterously slow, but with advanced molecular nanotechnology one might imagine the job being done without the cranium suffering the indignity of surgical rupture, and millions of artificial neurons might be switched every second. Rather than build them inside your head, a limited series of generic models might be floated in on the bloodstream, ready to have their menus set to map the individual neurons they replace.

It's worth noting, in passing, that if synthetic neurons can be made half the size of the organic varieties, we could even *double* the number of neurons inside your head in one hit. Would this automatically increase your brain power? Probably not, because specialized architecture is crucial to cognition. Still, one of the most palpable differences between modern humans and Lucy, the protohominid of the Ethiopian plains three million years ago, is that we own, on average, 1330 grams of brain tissue, while she had to make do with a third of that. With more components and some measure of plasticity in rewiring them, we might find ourselves becoming conspicuously cleverer in the ensuing weeks and years.

Leaving the brain behind

Despite these enticing (?) prospects, my purpose is not to sell you a brain upgrade but to carry the thought experiment a step further. Suppose, once the mapping and substitution is complete, that this detailed atlas of your brain is also copied into the huge memory and processing system of a supercomputer of the 2015 vintage—one easily handling as many teraflops or petaflops as a human brain—perhaps a hundred thousand trillion calculations each second.[5]

So this mindless, terribly fast machine now contains a digital description that emulates your original brain. If we arrange for streams of data from the outside world to enter its ports, just the same kinds as would normally enter your own ears and eyes and taste buds and movement sensors and internal monitors scattered through your organic body . . . what happens? For balance, let's add outgoing channels that permit the emulated brain to reach out into the world, touch and move objects, stroke and sniff and chew in synchrony with the incoming impulses that feed its sensorium. The simulated "you" will then, surely, feel himself or herself to "be" a "person"—to be, in fact, *you!*

Uploading yourself

Assuming we've left nothing out in this exercise in simulation, what we've achieved in our thought experiment is an *upload*—a complete dump of your mind into the flickering electronic whirr of a computer platform. Just to make sure you don't go mad at the shock of the transition, we will massage and morph the sensory data flowing in and out, ensuring that "you" genuinely *feel* a physical continuity with your old self.

This might require, to be brutally literal-minded, linking your cybermind to a humanoid robot extension replete with stereo TV cameras at the top, and two servo-mechanical arms hinged at elbow and shoulder, and five tactile, gripping fingers, and two prosthetic legs . . . Alternatively, your experience might be delivered, after considerable preprocessing, in the form of a convincing virtual reality construct. Only that small part of the world you're choosing to "look at" will "exist," but that portion will be rendered with the maximum available pixel-rich detail at the focal zone, fading away to impressions at the boundaries—just like now, in fact.

Is this uploaded personality conscious?

Is it you, or your twin, or something unprecedented?

If a disaster destroys the hardware it's running on, and the latest backup is reinstalled on a new machine, is the "new" version of "you" the same as the first, as you remain *you* when you sleep and wake? Or is it quite a different person, who just chances to recall everything that ever happened to you (well, at least with the fallible fidelity now available to your own organic brain)?

These are the kinds of quite shocking and exhilarating questions that will need to be settled as we ascend the slope into the Spike. Because, like it or not, the path into superintelligence could very possibly pass through upload territory before it reaches autonomous artificial minds.

Doubts about uploads

Before we get too excited, though, we should attend to the critics. One of them is Norman Molhant, at the Université de Montréal. Molhant doubts that this sort of upload is feasible, at least for the foreseeable future.

Here's his reasoning: the mind is not just a function of the architecture of the neurons, but is intricately modulated by the hormones that link them. Each neuron is itself already the equal of a minicomputer, coupled simultaneously to thousands of others in a network. We still can't fully simulate so much as a single neuron.

Moreover, unlike a living brain today's neural networks have fixed patterns of connection (although the "weightings" or values allocated to the different patterns they hold can be adjusted to reflect changes in experience). They have fewer links per neuron than human cortex. The whole package is locked to the outside world by hardwired input and output circuits. Can we expect to unravel the intricacies of a real brain within the next thirty years? Even if we manage to do that, in order to host an uploaded mind we'd still have to construct startlingly complex self-repairing computers with sophisticated links to the environment.

Are those objections sound, or is Molhant too conservative? Anders Sandberg, an amusing and brilliant Swedish transhumanist, replies that state-of-the-art neural networks have already improved on older models. Today's neural nets, he claims, "can reconfigure or emulate real, messy, biological neurons (actually, our current models create rather good results, albeit very limited in scale and still far from the fidelity we want in uploading)." Besides, while it's true that 100 billion neurons, each connected to many others by up to 10,000 synapses, is a whale of a lot of hardware to emulate, it may not be necessary to replicate everything in each neuron, down to the chemical details.

We might, of course, need to provide an equivalent for the diffuse washes of neurotransmitter chemicals that modulate much of the brain's emotional life, a feature often ignored in older attempts to create analogies between wet brains and dry computers. Software, needed to diagnose errors in the simulated brain and correct them, might be created largely independent of direct human effort (as we saw in chapter 3) by ecologies of genetic algorithms (GAs). Self-organizing strings of data that compete under Darwinian selection inside machine search-spaces, like the A-life creatures in Tierra, GAs can automatically compile good, if not absolutely perfect, solutions to well-defined problems.

Can all this be done within thirty years? "We tend to be surprised," Sandberg concludes evenhandedly, "at how fast or slow things develop; some enthusiasts claim we will experience a Vingean singularity in much shorter time, others are conservative and think nothing much will change. It is impossible to predict."

When machines run the world—a parable

We shall return to these consequences in subsequent chapters. For now, to get a sense of the quite drastic jumps that might follow quite swiftly upon such breakthroughs in either uploads or hyperintelligent machines, I'll share with you a charmingly grim sequence of scenarios proposed (not altogether in jest, I suspect) by an Australian defense scientist, David Bofinger. The moment we take the next step from uploaded personalities to self-enhancing uploads, an entirely new ecology bursts forth upon the face of the planet. This would be, undeniably, the Vingean Singularity, the Spike.

Bofinger, an applied mathematician, declares in advance: "I don't really believe in the extreme form of the singularity, since it's an attempt to extrapolate too far beyond the data that created the model. In other words, while I can't show why the trend of human history will eventually turn aside from the singularity, I think that Vinge's ideas will someday look as short-sighted as Malthus's ideas do today." Still, if it did happen, what dispensation might we expect these posthuman people to make of us, their predecessors? (Assuming some choose to be uploaded and enhanced in a nondestructive manner, while others remain in the utopian Limbo of a world with a new dominant species.) Will they tell us, as Yahveh did to Adam and Eve: you may eat of every tree in the garden, but stay off the grass? Here is Bofinger's merry little *conte cruel* of the Spiked future:

Monday: "You unmodified humans can have New Jersey, but stay out of the Everglades." Unmodified humans move to New Jersey.

Thursday: "Our predecessors didn't need New Jersey, but we do. We are moving all the unmodified humans to Antarctica." Unmodified humans walk through suddenly appearing arches and find themselves in an artificial environment on Antarctica.

Saturday: "The overmind has chosen to return Earth to its pristine state. You will all be relocated to a bubble-formed asteroid." Unmodified humans blink and find themselves somewhere else.

Sunday morning: "I have decided to . . . Oh, never mind, you wouldn't understand." Nothing important changes but there's some evidence the laws of physics have changed, or perhaps that everyone is now running as a computer simulation of themselves. No way to be sure, though.

Sunday afternoon: The speakers start to say something, then everyone ceases to exist. Maybe they were going to say, "Hi, we're all going to commit suicide now, and you have to go too." Or maybe, "Excuse me, but we need your raw materials." Or perhaps, "My God, how can you bear to live like that? Better put you out of your misery." Whatever they were going to say they got impatient with how long it would take and just went ahead and did it.

On a less apocalyptic scale, we are obliged to face the prospect of sharing the planet with entities whose interests and motives and activities we simply cannot comprehend. Is this true already?

In a sense, nations and corporations and religions and scientific disciplines are fuzzy self-sustaining systems with a certain integrity, striving toward goals not truly understood by any of the human atoms that comprise them. Certainly, in advanced computational mathematics, results are known (the solution to the four-color mapping problem, for example) that no single human mind could get across in a single lifetime.

Posthumans will be *different*

What will these indistinct but powerful entities, these post-Spike posthumans, be like, compared to us? Software developer Wayne Throop has conveyed this amusingly. While someone from the Renaissance would not comprehend a VCR or CD player, still, as Throop notes, in 1500 "there were (some sorts of) music boxen (methinks). Socially, you can explain a VCR to somebody from 1500 quickly and easily: 'See, you put this thing in here, ignore that flashing 12:00, push this here, and you get sound and picture out here.' "

By contrast, the Spike fetches *difference in kind* "Why did you go catatonic for a minute there?" "Well, I was . . . Let me put it this way, my link to what you used to call 'the Net' was . . . Um . . . well, I needed to . . . Um, well, you really got me there; I can't see any way to explain it to you in less than about hundred years of normal-speed conversation."

It's not, Throop adds, that such mysterious technology couldn't be explained on a superficial basis: "My forebrain was shut down while I uploaded my cognitive functions onto the Net for a while." What we on this side of the Spike would never be able to grasp is *why* that change of cognitive venue was deemed necessary, or what the posthuman was *doing* there. What vast issue was being considered that an unboosted human brain was insufficiently fast or complex enough to deal with? How did that elaborate issue relate to society's or even *that posthuman person's* goals? All of this, from first to last, might be too complicated to interpret to one of us unmodified merely-humans.

Should we throw up our hands, therefore, and wait for the Spike to steamroll over us? Or are such scenarios telling us something else, as Bofinger, for example, plainly intended—that the very notion of the Singularity contains some fatal flaw?

Doubting the Singularity

The more feverish grows the excitement of its proponents—and this does not include Vernor Vinge himself, who insists that the Singularity is simply a bound on our knowledge of the future— the more resistance is seen in people whom you might expect to be running out into the street beating a drum.

The transhumanists and extropians themselves have started to recoil from the gaudier proselytes. Anders Sandberg clears his throat with some embarrassment. There has been far too much Singularity-worship, he observes. Is the Spike inevitable, easy, Destiny itself? No. "It is a neat idea invented by Vinge that has caught our imagination and fits in with *some* trends we see. There is, in my humble opinion, no more real support of it than the inevitability of the dictatorship of the proletariat."

He was not alone. Extropian director Dr. Max More was scathing: "The Singularity idea has worried me for years—it's a classic religious, Christian-style, end-of-the-world concept that appeals to people in western cultures deeply. It's also mostly nonsense . . . The Singularity concept has all the earmarks of an idea that can lead to cultishness, and passivity. There's a tremendous amount of hard work to be done, and intellectually masturbating about a supposed Singularity is not going to get us anywhere." And Augustine stated flatly: "I think the 'Singularity' is a myth, an illusion."

It was quite shocking, really. Here were people whose lives are organized around the prospect of (scientific) immortality, of extravagant change, of—in More's words—"a wonderful posthuman future where we will be unshackled from many human limits." But it would have to be *created*, not merely endured or submitted to.

In fact, nobody seriously looking up the soaring curve of the Spike doubted this. Certain New Age doctrines might smack of this supine god-struck reverence—those of the late psychedelic guru Terence McKenna, say, who perceived a "novelty curve" Timewave supposedly derived from the *I Ching* that would peak on 21 December 2012. Others saw clearly enough that the exponential curves mapping the likely upward sweep of technology was nothing else than the sum of human efforts in the next twenty or fifty years, aided by the fluent and growing power of machine amplifiers.

Still, some observers were clearly worried that this temptation to passivity and worship might corrode the very impulse required to make it happen. Little value in paying good money to have your head frozen after death, in the hope that nanotechnology would allow future generations to revive you, if everyone went on an extended holiday, lolling about in the expectation of the Singularity arriving miraculously.

What was needed was hard thinking about the details of life during the Spike. What would it really be like, for example, to have your mind uploaded into a computer? How would it feel to have your intelligence amplified, or to live unmodified in a world where other people had already taken that path? Let us turn to those questions, to the texture of living in a world moving into the Spike.

6: Uncoupling the Flesh

THE FUTURE IS HARD TO PREDICT. WE MAY FEEL CONFIDENT THAT
EVENTUALLY SPACE WILL BE COLONIZED, OR THAT EVENTUALLY WE'LL
MAKE STUFF BY PUTTING EACH ATOM JUST WHERE WE WANT IT. BUT SO
MANY OTHER CHANGES MAY HAPPEN BEFORE AND DURING THOSE
CHANGES THAT IT IS HARD TO SAY WITH MUCH CONFIDENCE HOW
SPACE TRAVEL OR NANOTECHNOLOGY MAY AFFECT THE ORDINARY
PERSON. OUR VISION SEEMS TO FADE INTO A FOG OF POSSIBILITIES.
—ROBIN HANSON, "IF UPLOADS COME FIRST" 1994[1]

Can we sensibly plan for direct linkages between mind and machine? Rudimentary EEG-controlled switches have already existed
for years, using 10 microvolt alpha signals from the temporal and
occipital lobes.[2] So have rather primitive data-channel prostheses
for the deaf and blind.

For example, Richard Normann and colleagues at the University of Utah are working on a neural implant able to pipe 1024
pixels from a small head-mounted camera, via a "transcranial
interconnect," into the visual cortex of a blind person. The resulting image is grainy but useful. Harvard Medical School's Jo-
seph Rizzo and John Wyatt, an electrical engineer from MIT, have
cofounded the Retinal Implant Project, planning a photodiode
array inside the eye itself to help those whose retinal rods and
cones are damaged by macular degeneration. In January 2000,
Dr. William Dobelle announced that a computer-aided system
already helps a blind man known only as Jerry to navigate using
just a hundred pixels that mark the outlines of objects his camera
and ultrasound sensor identify. The system is still primitive, ac-
tivating 68 platinum electrodes implanted in his brain in 1978.[3]

More impressively, in November 1999, Californian scientists
announced that they had intercepted the neural transmissions of

a cat's eye from *behind* its optic nerve (where its eye had already done some preprocessing). Using "linear decoding technology," they'd translated these brain waves into grainy but identifiable images.[4] Monitors displayed what the cat was seeing. Once such technology becomes practicable for routine human use, several extreme possibilities will open up.

We might see Dan Clemmensen's Trans Web crystallize into existence, a kind of dispersed and conscious Internet group-mind blending people and a global computer net. Or the machine end of the jerry-rigged composite might learn from its human adjuncts how to think for itself, in a deeper, accelerated version of the process by which a computerized neural net is now "trained." These useful gadgets already learn to produce desired outcomes through trial and error, without anyone really knowing in detail how their hidden black boxes are rewiring themselves.

Most confronting of all is the upload option. A brain might be scanned in exquisite detail, and its very structure replicated inside an early twenty-first-century petaflop computer. Unless doubters such as philosopher John Searle and mathematician Roger Penrose are correct, its contents would be replicated also, and its states of mind. What then? "If we could make exact copies of someone (not just clones, but exact down to quantum limits) what would this do to our concept of ego?" Vernor Vinge has asked. "This is just one (and one of the simplest) of the problems that I see looming in our future. Our most basic beliefs—the concept of self itself—are in for rough times."[5]

Destructive scanning

Suspended animation advocate Paul Wakfer points to one drawback of plausible uploading. A former professor of mathematics and physics, Wakfer headed the privately funded Prometheus Project, now defunct (or at least suspended). It aimed, unlike cryonics, to achieve reversible suspended animation within twenty years, through a program of directed research. Wakfer still hopes to achieve resuscitation from the preserved state *without* destructive dabbling with the temporary corpse. Critiquing the upload path to preservation, he notes that "even after the development of computer speed, size, complexity, and software able to handle uploading (many decades away at best, in my opinion) the most likely approach to complete mental readout is destructive scanning of your undamaged frozen brain."

In other words, sending an infestation of nano copiers into your living brain is a less likely near-term option than having your brain frozen immediately after death (perfused by cryopreservatives to minimize damage to the delicate tissues), and then slicing it into extremely thin wafers for analysis. "Slicing" might be too crudely mechanical a term. Perhaps a destructive scanner might very cautiously map the brain with a laser beam, evaporating the web of neurons as it goes.

At any rate, the information would be transferred to a massive machine memory. There, after a reconstructive protocol we can't yet begin to specify, your informational double would awaken in some sort of virtual world, linked to ours by sensors and hence complete with convincingly sensuous experiences—ready to enjoy the company of others who had gone before you. It would be the "white light" tunnel of the Near Death Experience, except that now it would be real instead of hallucination.

But would it *really* be real?

Is a copy living as an emulation inside a digital space in any authentic sense *real*?

Why not? This is not a mystical proposition. I am not suggesting that your "immaterial soul'—whatever that might be— will be grafted across, to attain reincarnation inside a silicon central processing unit. Rather, it's as if a perfect copy of you appeared, if not down to the last quantum then at least in some neurological detail. That kind of fidelity would make the digital copy a real person, a true human being with consciousness and volition and desires. He or she would be no less a person, despite the obvious lack of a protein-based body, than is cosmologist

Stephen Hawking, his brilliant mind constrained by disease-damaged and failing flesh.

Even that is a misleading way to put it, for it risks misinterpretation. The fleshy body is not a prison for some impalpable mind-stuff. That is the error of dualism; philosophy and science alike have taken centuries to escape it. Mind *is* flesh, we know of no other kind. One day, other varieties of minds will have the same relationship to other kinds of substrate—that is, they will be *what is generated* when the appropriate enormous interactions occur within a suitably structured ensemble of atoms. Call that process a "computation," if you like, or just say that systematic changes flow back and forth through the substrate, so that its final state remembers what has happened during its experiences— *remembers*, is *aware* of the world and of itself.

Let us stipulate this much without further ado, for it would be futile to argue over basic principles. Some simply will not countenance the claim that a machine can serve as the site for a consciousness in the same way that a brain-body can. So-called "mysterians" like British philosopher Colin McGinn insist that there is something about a mind that might not be duplicated in a machine, perhaps something separate from the brain itself: "Neural transmissions just seem like the wrong kind of materials to bring consciousness into the world." David Chalmers, professor of philosophy at the University of Arizona and associate director of the Center for Consciousness Studies, and certain other young philosophers, have lately revised this sort of enduring doubt into the status of an intellectual miniscandal.[6] They speak of the "hard problem" of consciousness—the difficulty of seeing how *qualities* of experience (*qualia*) can be derived from information flows, however ornate, inside a bunch of organized atoms. Others, like Daniel Dennett, and Paul and Patricia Churchland, see no difficulty in an appropriately reductionist stance. I will provisionally accept the view that allows us to keep asking scientific questions. That is, the view that permits the use of systematic *doubt* coupled with rigorous factual *inquiry:* the unfinished materialist account of cognitive science.

Is it still you?

Consciousness, Chalmers considers, *supervenes upon* the material structures of the brain, and is not identical to it; he is a kind of dualist. Yet, perhaps surprisingly, Chalmers does not at all dispute the likelihood of machine intelligence. Indeed, he writes: "I will take things further and argue that the ambitions of artificial intelligence are reasonable ... not just that implementing the right computation suffices for consciousness, but that implementing the right computation suffices for rich conscious experience like our own ... It might be ... that a computation that mirrors the causal organization of the brain at a much coarser level [than a neuron-by-neuron emulation] could still capture what is relevant for the emergence of conscious experience."[7]

But even if we accept the arguments of those who regard the mind as something that grows as the brain and body grow (and can in principle be replicated with a high degree of accuracy), it does not follow that having your brain mapped into a computer is the key to endless life *for you*. Upload enthusiasts commit a central fallacy, one absolutely crucial if you're planning to shuffle off this mortal coil in exchange for life as a supercomputer.

They promote their ambitious vision in a number of ways, all flawed (if I'm right about this). Once you've been uploaded into a machine, they claim, you have a kind of *security* never previously possible in all the history of complex life on this planet. A simple unicellular organism has the power to replicate its strand of genetic material, pump up in size, and split into two or more copies. With the coming of sex, and its pest-thwarting, immunology-enhancing gene recombinations in the offspring, that path to unlimited longevity was closed.

It need not have been, but the energetics of mindless evolution ensure that disposable bodies carrying remixed genomes beat competitors for survival of genes, at least with large, complex creatures. So we lost immortality. Scientific tinkering might give it back, perhaps by engineering telomerase and other repair systems, just as it now gives older people back their failing eyesight using lens surgery or simple spectacles.

Consider a related example from the advancing edge of biological science. While laboratory cloning has returned to us that ancient option of faithful self-copying, a brood of quintuplets genetically identical to yourself as a newborn baby (which is really all a clone is) would hardly *be* you.

Why not? Because environment and the fine-grain detail of experience makes us, as much as genetic inheritance. The clone quins' memories would not be yours, even if you struggled to give them closely similar experiences. After you died, they might feel a special pang of loss—or they might be bitterly grateful to be rid of the interfering old buzzard—but *you* would be *gone*.

Selfish genes might act as if they have an urge to multiply themselves, and indeed nothing is quite as effective in that (in the short term) as literally replicating your genome nucleotide by nucleotide for the whole hundred thousand genes. But still, I insist, the copies *are not you*.

Why should it be any different for the computer-uploaded copies of your brain? Well, comes the obvious answer, because they actually *do* have your memories. Full backups can be kept, just in case the power fails and your memory file vanishes into digital dust. Or, worse, when some fiend steals the computer and reformats you out of existence, writing a virtual reality game over the top of your bits and bytes. Convenient to have some flunky haul out the backup and reinstall you on another machine. Or, to take the long perspective, when the sun expands into a red giant in several billion years' time and evaporates your hardware along with your planet. Nice to be able to send off, in the nick of time, a hundred or a million nanostored copies of yourself, fired or beamed into the safety of deep space.

All these prospects, however far-fetched they might seem, are perfectly feasible. At least, there's nothing logically wrong with the idea. Except—

The exact copies still are not you![8]

How do you tell a perfect copy from the original?

Actually, in a perfectly valid sense they *are* you—but how does that do *you*, here and now, any good? Speak to one of these perfect copies, and she or he will be certain that his or her identity of self has continued unbroken since the scanning was done. (Or a little earlier, assuming your original brain has to be killed or at least disassembled during a destructive upload.) There's a tear-jerking moment in that great old Kirk Douglas epic where first one and then another of the guerilla gladiator's loyal but defeated followers rise to shout defiantly at the Roman foe: "I'm Spartacus." "No, *I'm* Spartacus!" "Over here!" In the upload future, your copies will each yell your name, and they won't just be pretending out of devotion and fealty. They'll be speaking the simple truth—as each of them *experiences* it.

True, as time passes and their experiences differ, their identities will diverge. The *you* who wins a Lotto jackpot and retires to a huge domed estate of Mars will have a different life from the *you* who stays home looking after the family, let alone from the *you* who was kidnapped and put to work by fundamentalist cyber-slavers.

Would you die for your upload double?

The piercing question is this: would you be prepared to terminate your own stream of awareness, just so that some other person (with the same memories, admittedly) could awaken to an adventure in virtual reality, to endless machine life?

Some say yes: after all, once you're inside that computer, the benefits never stop. No more colds or cancer. Your mind might be able to run at dizzying speed. No longer restricted to the sluggish baton-passing of neurotransmitters and ionic currents, your electronic or optical consciousness stream would blaze like the pure spirit of some Miltonic angel. And never forget those convenient backups, and the security they represent.

Frankly, I find this kind of reasoning utterly baffling, even though it's undeniable that you and I go happily into sleep (and perhaps less happily into medical unconsciousness) fully confident that the "reconstituted" self that later will wake is entirely continuous with the present you or me. And of course the only sensible motive for arranging to get one's head frozen at death would be the conviction that a revived or uploaded brain will be just as much "me" as "I" am after a snooze. But really, *destructive* emulation—

Why should I care about *his* greater wealth? *Their* security of tenure on a dangerous planet?

Selfish genes meet the Spike

Evolutionary sieves have winnowed our genomes in favor of building bodies whose economic behavior is channeled by the need to sustain (and ideally reproduce) the components of that genotype—the individual selfish genes, and the slightly less selfish ensembles of genes that do well together.

According to a standard argument developed by mathematician Ronald Fisher and geneticist William Hamilton, we will tend to be "altruistic" toward other bearers of large chunks of the same genotype, because in a genetic sense They-R-Us: our children, our parents, our siblings, even our cousins.

Since evolution is mindless, this mechanism allows high-level adapted structures such as brains and cultures to make "mistaken" identifications. Hence, individuals can sacrifice themselves in support of the "wrong" genotypes. Young men hurl themselves into the firing line because, in a sense, their genes have been tricked. Among other reasons, they have been persuaded to bond with their (genetically distant) fellow warriors, with their nation or religion. Still, it's clear that the selfish-gene altruism equations would be satisfied if I were to sacrifice my body in order to produce a dozen copies of my exact genome, with or without cultural and individual memories. *I*, however, would not be so ardent. Trust me.

Suppose I could arrange for a dozen exact copies of my body

by cloning: a time-lapsed set of identical-tuples. My genes would
rejoice, if they had the brains to understand what was going on,
but if I could only achieve this by giving up my individual life I
can assure you I would not share the joy.

Exact copies

Suppose, however, that these copies could also contain my exact
memories to this moment, as we've stipulated above, so that I
spawned a dozen true copies of myself. (These immediately
would split off from each other, it's admitted, in terms of experience, random wiring events, future identity in short.) Would
this offer make me more inclined to die in order to achieve perfect duplication? Certainly not. Why should I care about *their*
enhanced prospects?

To spell this out quite clearly: if I were offered the chance of
a lethal injection in exchange for the creation of a perfect (or
even renovated) version of my phenotype, or ten or a million
copies for that matter, I would mutter a very rude word and go
home.

But suppose I were dying of a currently incurable disorder,
and a cryonics company offered an effective postmortem neurological suspension attainable only if I were temporarily killed
in the process (if it were known, say, that waiting for death by
senility or disease caused the loss of too many brain cells to permit resurrection). It doesn't even matter, for this question to apply its sting, whether I am offered a new, young body grown from
my stem cells and imprinted with all my memories, a robot
equivalent, or an emulated and superior cyber version inside a
high-petaflop machine. Would I go gentle and immediately into
that good night? I don't know. These are *difficult* issues, and the
answers not at all self-evident.

Indeed, it might not be necessary for me to be at immediate
risk of dying, except in the sense that we all are. Perhaps the
sooner you get your brain destructively scanned the more reliable
the result will be. What would *you* do?

Learning to be you

Everything depends upon what I mean by "I," what you mean by "I." Which is what's at stake in this apparently remote and abstract matter of identity. Greg Egan has proposed a series of brilliant "thought experiments" dealing with identity and its duplication, playing with these questions in a series of skin-crawling *noir* narratives. One favored mode of upload/duplication/extended noncarbon life is a growing "crystal" implanted in the brain, that echoes (in effect, mirrors or redundantly stores) every brain state from then on: "learns to be me." It's a bit like the two cerebral hemispheres, to the extent (whatever that extent is) that left brain and right brain are not complementary but redundant backups. Eventually, when your consciousness is running on a joint system, neural net perfectly echoed by the crystal, your meaty brain is . . . scooped out, and thrown away.

To the unprejudiced, this apparently nightmarish operation should be no cause for alarm. Had the mirroring been done at a distance, onto a human-capable machine via radio, the same situation would obtain. So too if our earlier nanomachines had constructed the copy. (Egan's narratives cleverly find the weak fracture planes in such proposals. What happens, for example, if crystal and brain are accidentally decoupled before the meaty brain is destroyed. Which one then is "me?") It *is* alarming, of course. It is terrifying.

Can we find some case where duplication has no distressing consequences? Suppose all twins were born perfectly telepathic, so that each experienced what the other did. If one of the two died, would that *really* matter? You might grieve today if one of your arms were amputated, but you wouldn't think that you'd thereby lost one of your selves . . . That analogy doesn't quite work, because even telepathic twins are differently situated *as visible persons* in space, and the loss of one body would surely devastate the other locus of their shared consciousness. Still, it undermines simple views about identity.

Well, what if everyone were *given* a cloned double, with whose brain states he or she slowly became redundantly resonant . . .

Sorry, it doesn't work. It's hard to retain empathy for the subjects of these thought games. Vernor Vinge, who's spent a great deal of time on these problems of the Spike, went some way toward exploring the experience of a shifting group-mind in his novel *A Fire Upon the Deep*. In such a disseminated consciousness, made up of wolflike individuals, some modules can die and get replaced by different individuals entirely. The character of the ensemble would alter, but a core of identity of the group-mind/bodies might prevail. Or so Vinge claims.

The bottom line, I conclude, is that physical and general brain-process continuity support our current sense of continuing identity. They allow us to believe that the person who wakes up tomorrow is the same one who went to sleep tonight. What of awakening after ten years in a coma? Nobody denies that this case is deeply traumatic, but still we assume that *the same person* has woken up. Waking after half the brain is removed to forestall death by cancer? No one denies that this case is even more deeply troubling.

But look—Even if *I* can't tell the difference between you and your perfect clone, *you* surely can, as they strip your brain down.

Don't beam me up, Scotty

If I were copied by a teleport machine, Xeroxed elsewhere by a "transporter beam," and the one at the transmitter end was to be killed (to even up the metaphysical balance of the universe),[9] I'd go kicking and screaming . . . I'd be not one whit reconciled by the simultaneous existence of my double in the receiving pod—*unless* I had remained, throughout, in perfect nonlocal continuous connection/identity with him. In that case, I'd swiftly get used to the notion that it was just like dropping off to sleep on the jet and waking up in another town. Of course, I could be *fooled* into believing that this were the case, even if a destructive-uploading expert knew it wasn't, which would soothe my dread. "Just step into the nice warm shower, Damien . . ." But I'd rather not be deceived in that way.

Would you be prepared to die (sacrifice your current embodiment) in order that an exact copy of yourself be reconstituted elsewhere, or on a different substrate? To insist upon this question is not to be a hostile "upload skeptic," at least not in the sense that the philosopher John Searle might be, denying that a consciousness could ever be implemented on a computer system.

Others disagree. "Are you so positive," someone asked Pelagius, the extropian nanotechnology enthusiast, "that you would flash upload this instant, given the chance?"

His answer amazed me.

"YES!" he said. "In a heartbeat."

Yet it would be *his* very last heartbeat . . . Pelagius's argument is that when you examine two physical systems and find no difference between them, there is indeed no difference. He claims as "excellent evidence that I am still alive after I am fast-uploaded" that "I produce something that has my memories, acts just like me, and insists that he is indeed me. As good as this evidence is," he adds, "I admit it falls short of a proof. You will *never* be able to prove that my upload, fast or slow, has my consciousness, but then, you can't even prove I had consciousness before the upload, just as I can *never* prove you are conscious. Nobody worries about being the only conscious entity in the universe because in our heart everyone takes it as an axiom of existence that when something acts intelligently it is conscious. For the same reason, I don't worry about an upload who acts just like me not being me."

The two human species: flesh and upload

In a sense, then, the philosophical question is moot. As long as there are people who share this blithe conviction that identity persists through upload, they'll buy the service the moment it's technically feasible. Unless the machine emulations of Pelagius and his friends begin to bemoan the error of their choice (and why should they, since they share the memories and disposition of their predecessors?),[10] the conscious inhabitants of this planet

will be split into two species: humans in their original and continuing bodies, and uploaded copies of dead humans. Add to that
a third category, the independent minds built from the ground
up by artificial intelligence programs. And perhaps a fourth: living, embodied humans in an extended condition of enhancement, plugged in (perhaps only some of the time) to the
super-Net.

What happens in a world like this? How do we deal with such
a proliferation of new intelligent species? The dawn of that age
will precede the Spike proper, I think, because its transhuman
condition—only verging upon the truly posthuman—does not
escape the power of our educated imagination. Even intelligent
machines can be understood, and probably it will be initially important that we make the effort to do so.

Cautions about upload and AI prospects

The neurophysiologist William Calvin looked into this question
when Vinge's paper on the coming Singularity was published in
Whole Earth Review in 1993.[11] Calvin stressed that the impact of
machine intelligence, assuming an AI can talk and even chat like
a human being, will be profound.

Early attempts to program such a machine will lack the fine-
grain detail of human behavior, but even a device that simply
works *somewhat* like a human mind "will set in motion one of
those historical transitions," he acknowledged, "after which nothing is the same. Perhaps it won't qualify as a singularity (an instant shift into totally unpredictable consequences) but we surely
have a major transition coming up in the next several generations
of humankind . . ."

Calvin's expert opinion comes down against uploading, shunt-
ing a living mind into a computer, more for psychological than
engineering reasons but for those as well. "I suspect that [it] is
unlikely to work; dementia, psychosis, and seizures are all too
likely." In the meantime, we can learn a great deal from the challenge of programming an AI "workalike" to behave like a

person—a device that "reasons, categorizes, and understands speech" and is "as endearing as our pets."

That last characteristic is something missed by many mass-media portraits of humanoid computers and robots, which used to be depicted as cold, rational monsters (who nonetheless managed to bungle their evil plots in bizarre fits of irrationality). George Lucas changed all that with the cute bleeping robot R2-D2 and its bumbling droid sidekick C-3PO, and Steven Spielberg's lovable E.T. probably undercut a whole generation's xenophobia for minds and bodies different from our own. Even Arnie as the Terminator, ruthless and nightmarish, was tamed into the ideal Dad ("No problemo!") in the sequel.

Ordinary hackers are adroit at personalizing their home computers, giving them names—as people have done for decades with their idiosyncratic cars—and filling their screens with friendly winged toasters, roaring tigers and dragons, and messages either in script or simulated speech: "Good morning, Damien, this is Igor. You're looking especially masterful today, sir!" Or: "I'm afraid the Cray is down, my lord. You'll be driving with me today." Or: "What a *wonderful* life!" Or even: "*You* again. Bah, humbug."

Artificial intelligence, Calvin suggests, will be replete with everything a self-reflexive consciousness requires. It will focus its attention on a given task, finding its way to solutions by way of imagery, contingency planning, and narratives about the world and its place in it. Literally, it will tell itself stories and trace out their consequences. Not all of this processing will occur at the same topmost level of awareness—indeed, the bulk of it will be done, as ours is, in hushed, partitioned regions of the artificial brain. The key difference is not that human thought is largely unconscious, but that machine intelligence could be blindingly faster than ours. This implies, Calvin tells us, that "we'll see an aspect of 'superhuman' emerging from the 'workalike.' "

AIs and Darwinism

Even before superintelligence makes the world more tense for ordinary people than it is now—and this could occur due to genetic engineering, as well as AI or uploading—we will find ourselves in a disrupted psychic ecology. Species competition, Calvin reminds us, is fiercest between relatives. We do not usually fight with birds and bees for living space. The soil continues to be churned by uncaring worms whoever walks upon its surface. Our ancestors, however, did wipe out all the other primates that got anywhere near our ecological zone. Of our omnivorous cousins, only the chimpanzees and bonobos remain, and they persist at our sufferance. True workalike AIs (and uploads) will suffer the same Darwinian neighborhood pressures, as will we.

Most of the time, humans no longer grub for, well, grubs; our income derives from playing abstract roles by abstract rules in fabulously complex economic networks. Food production has dropped back from the chief toil of First World humankind to the machine-reliant business of just three percent of our number.

When human-level machines replace people in many of the more cognitively demanding occupations, as today's computers have mown down the ranks of typists and those who added and subtracted for a living, we'll see a fight for survival. Already one encounters neo-Luddites calling for the end of machine civilization. But benefits flow as well: a great deal of back-breaking toil has been abolished already, and so will much of the boring intellectual grind that stifles the imagination and wearies the brain. We just have to ensure that the dividend from this terminal downsizing of the human economy is shared among those dispossessed through no fault of their own. It is a topic we'll return to in chapter 7.

Meanwhile, new advantages are on the way. Today's CD-ROM language programs and mathematics courses can teach tirelessly, at each student's own pace. Smart machines will make education ever more responsive, tailored to the individual, interesting and interactive. Of course, this assumes you can have "smart" machines that are not also bored or enslaved and rebellious minds-in-silico. Uploaded human minds, unless they are savagely edited, will own as much hunger for self-determination as any other. No doubt criminals and wicked political regimes will literally enslave some minds in boxes, punishing them with inescapable pain and rewarding them with the equivalent of nonstop orgasm or opiate highs.

We are confronted, in a way, with the poignant dilemma of any attempt to explain the ways of a god to a human. If we build human-level machines, let alone uploading our own minds into such platforms, we create beings with interests and desires and fears and vulnerabilities. Those beings will not be simple duplicates of us, so their crimes and temptations may differ from ours, but they too will know the Problem of Pain and perhaps the Problem of Evil.

The values of a smart machine

How are we going to instill our own preferred values into intelligent machines? Can you teach a machine to be moral?

Calvin claims that AIs will start out "totally amoral, just raw intelligence and language ability. They won't even come with the inherited qualities that make our pets safe to be around." While we can program AIs to look and act superficially cute, they will lack the long coevolutionary history that domestic beasts of burden and pets share with *Homo sapiens*. The solution might be a serial breeding program: we can digitally "clone" the versions we approve of, adding slight variations, and keep the ones that do best. Thus, we'd be emulating Darwinian selection processes at a vastly more rapid rate.

In another sense, of course, we'd be introducing Lamarckian evolution instead (the inheritance of acquired characteristics). Each fresh generation will retain the experiences of its ancestors as well as their blueprint. "The early models could be smart and talkative without being cautious or wise, a very risky combination, potentially sociopathic," Calvin warns.

Just as governments now license and control the use of dan-

gerous pharmaceuticals and biological agents such as cloned and
genetically engineered food stocks, it might prove desirable to
contain new AIs and their discoveries. "There might be a one-
day delay rule for distributing output from superhumans that
only had a beginner's license," Calvin suggests, by no means
whimsically, "to address some of the 'program trading' hazards."
In extreme cases, we will require the computer equivalent of a
top-level containment laboratory, the sort used today when replicating lethal viruses for study.

Caging the AI tiger

Containment might be a lost cause, however. Machines may pass
swiftly from human-level to superhuman intelligence, their improvements made possible by exactly the evolutionary sequence
mentioned already, where star pupils from an initial variety of
programs are trained in different ways, and the best of them
cloned, and so on for many very fast generations.

Once that transition has happened, there will be no retreat, no
postponing the conflict between humans and smart machines. We
need to think this out in great detail right now, while it is still
only in prospect. Our goal will be to create a world where all
kinds of conscious entities have a shared stake, where unenhanced
people (perhaps the vast majority, but not necessarily for long)
and artificial minds and uploaded individuals and emergent "colony minds" can participate in a genuinely synergetic ecology, each
helping to sustain and perhaps bootstrap all the others.

Calvin is relentless, and I do not find any fault in his reasoning.
The end result of these changes will not be a world just a little
different from today's. It will be the Spike: "Our civilization will,
of course, be 'playing God' in an ultimate sense of the phrase:
evolving a greater intelligence than currently exists on earth. It
behooves us to be a considerate creator, wise to the world and
its fragile nature, sensitive to the need for stable footings that
prevent backsliding. Or collapse."

The difference from traditional cautionary tales of "playing god" is that we shall have created our betters. If the uploading project makes sense, we shall also *become* our betters.

AIs or uploads?

A major obstacle to near-term artificial intelligence, as I've mentioned previously, is the sheer difficulty of writing suitable software. We still don't know how to turn a human brain into a logical flowchart, despite thirty or forty years of optimism from the AI and cognitive science communities.

One reason has been illuminated by Gerald Edelman and William Calvin and their supporters, who've shown how much of any given brain's circuitry is custom-built. Your brain is a statistical assemblage that has literally *grown* into a functional shape from the few broad hints laid down in a mere hundred thousand embryonic genes. Neurons make their connections in a somewhat blind, blundering way. We start with far more than we need, and experience prunes them back, permitting a huge amount of fine-tuning from whatever strange environment this particular little creature happens to be born into. Even when tropisms—tendencies to act and react in certain stereotyped ways—are "hardwired," fixed from an early age, still they are far more plastic than an old-fashioned neurological determinist might have hoped.

So, while the operation of the brain can validly be described as computational, the software is not easily distinguished from the hardware. If the grammarians influenced by the theories of Noam Chomsky are correct, our capacity to use language is built in to the structure of the brain. Cognitive scientists go further, locating language specialization in certain neural modules such as Wernicke's area in the left superior temporal gyrus and Broca's area in the left third frontal convolution. Is this programming or mechanism? Both, surely. Yet the precise shape our linguistic skills take is a matter of individual development, as we respond to everything we hear and see and especially the other people we meet.

Every child babbles a stream of random sounds that contains

all the possible phonemes utterable by the human larynx, tongue, and lips. Swiftly, because parents and other children love to talk back to tiny children, this stream of primordial gibberish is thinned and trimmed, shaped to the forms of local speech patterns. Later, as our vocabularies expand and deepen with specialized vernaculars—the exact technical words plumbers use, or brain surgeons, or accountants—the process starts to resemble changes in software rather than hardware. As we learn the kinds of logic and rules of inference preferred by our community, that too is new, acquired software. Yet it becomes anchored into the physical shape of the brain, and can be compromised or degraded in strangely specific ways if a very localized lesion damages a specialized neural module.

Reprogramming in space

Silicon computers, naturally, are far more readily broken apart into coding versus hardware. Failures of onboard systems meant that the 1977 *Voyager* space probe had to be tweaked by radio many years later, as it fled toward the outskirts of the solar system. That this was possible at all is due to the happy circumstance that an old, rudimentary computer—and recall that this mission was launched at the dawn of the age of miniaturization—can be reprogrammed "on the fly," its hardware upgraded with new, efficient software sent from home.

Similarly, something went badly wrong with the 1989 *Galileo* spaceprobe as it approached Jupiter, preventing the High Gain Antenna from working. Everything else had gone superbly—and there can be eighteen microcomputers running on *Galileo* at any given time—unlike other recent expensive missions that simply vanished into the endless dark. A NASA briefing notes: "It is important to remember that at the time Galileo was being developed, spacecraft designers were using fairly state-of-the-art computers, but those computers are extremely slow by contemporary standards."

Galileo slid safely into orbit, zipping by several of the major moons, but its fabulously valuable data had to trickle back to Earth through ancillary antennae. Even this trickle has been salvaged only because of new data compression software upgrades sent to the distant spacecraft, second thoughts broadcast from Earth. NASA adds, "Software has evolved from what was originally launched with the spacecraft in 1989—[and] improvements really *are* better! The software will continue to evolve for the remainder of the mission." Seven years after launch, as the probe neared Europa, "new software commands were successfully radioed to the spacecraft to further improve certain recording operations." Such on-the-run software upgrades are now routine with robot probe missions to the boundaries of the solar system.

Scanning the living brain

It could turn out, though, that emulating a living human brain, consciousness and all, is very much harder than AI enthusiasts hope. Agreed, Moore's law will probably keep tracking the inexorable improvements in computer hardware: size shrinking, numbers of components expanding and performance roaring ahead (Intel's chairman and former CEO Andy Grove, *Time*'s 1997 "Man of the Year," predicted that by 2011 we should see chips with a billion transistors, running at ten billion hertz and processing a hundred billion instructions per second). Yes, self-bootstrapping will help program these wonderful machines. Yes, cellular automata and genetic algorithms will "breed" software solutions by Darwinian competition, without the need for human ingenuity to guide them. Still, some experts continue to doubt that we will have the programming prowess in time to take advantage of the raw grunt. All dressed up, and no way to think.

That doesn't mean we can't build an artificial mind. In a cool, clear-eyed analysis of these prospects, former AI researcher Robin Hanson (now an economics professor) concludes that it might well be quicker to pull a real brain apart, in much the way described in previous chapters, and replicate its procedures inside one of these hypercomputers. This, after all, is the human upload solution.

Not every atomic or even chemical detail of the brain's wet cells is required, just enough to emulate the flow of data from point to point in its architecture. Such an upload, Hanson explains, would be software with human-level intelligence, "yet created using little understanding of how the brain works, on anything but the lowest levels of organization. In software terminology, this is like 'porting' software to a new language or platform, rather than rewriting a new version from scratch (more the A.I. approach). One can port software without understanding it, if one understands the language it was written in."

Suppose such uploads become feasible sometime in the twenty-first century, before genuine artificial intelligence is developed. What effect would this have upon society? Must we instantly plunge into a Singularity that is bound to be so strange that we dare not speculate about it? Not at all. Economic laws will remain in place, Hanson insists. The marketplace will adjust. But things certainly will not go on as they are.

Brains for hire

Sitting here at my word processor, I am already a cyborg, of a rudimentary kind. The machine is an amplifier of my abilities. It checks my spelling, brings me cogent information from the global Web, allows me to utter half-baked half-sentences and then erase or modify them without pain. A spreadsheet program does even more for anyone who needs to model financial and other complicated numerical flows.

Take the next step: suppose another human intelligence could be loaded into your desktop machine (or hired from the Web). Assuming this uploaded person was prepared to do the work, you'd have a genuinely intelligent and ferociously fast support system at your disposal. (I have to say that even putting the possibility into words like this is quite disturbing, with all kinds of overtones of involuntary servitude and solitary confinement. Let us trust that these qualms can be answered. If they can't, this is the worst nightmare anyone ever suggested.)

Hanson is somewhat blithe on this score: even with crude sensors and actuators, "uploads might not only find life worth living but become productive workers in trades where crude interaction can be good enough, such as writing novels, doing math, etc. And with more advanced android bodies or virtual reality, uploads might eventually become productive in most trades, and miss their original bodies much less." Obviously novelists and mathematicians are a sorry lot who live for their rather abstract work and rarely see the light of day. He adds: "Thus some people should be willing to become uploads, even if their old brains were destroyed in the process. And since, without A.I., uploads should be productive workers, there should be big money to be made in funding the creation of such uploads."

Life styles of the rich and uploaded

The logic here is just as elusive as Pelagius's. I assume that Hanson will gladly take the plunge and have his brain evaporated if there's plenty of money in it for his uploaded copy, or at least a secure job. The really big money will be made, however, by the companies developing upload technology. It is by no means clear to me that the CEO and programmers of Upload International will happily make the destructive-upload move to silicon Vale. Leave that to the brave and death-driven.

Let us allow Hanson this premise, though, and see where his logic leads us—once granted that *some* people will indeed agree to postmortem life as uploads.

An unusual feature of Hanson's scenario (one shared with robot designer Hans Moravec's) is that uploads might not be restricted to life in digital virtual reality. Synthetic bodies, produced by nanotechnology, can either contain very small, very powerful Drexlerian supercomputers preloaded with your mind, or serve as remote-controlled vehicles for your point of view. Either way, brains uploaded into computers will run at whatever clock speed is deemed desirable or affordable. As now, you might desire a supercomputer but only be able to pay for the future equivalent of a creaky old Celeron running at a sluggish 466 MHz. The faster and most commodious of these upload platform, Hanson notes, will make their inhabitants veritable "gods." Expect to see social hierarchies based on such hardware factors. Upload life will not necessarily be a blissful merging into a group colony mind; far from it.

No, it's dog eat dog in there, and more important, dog *beget* dog. If you can be coded and copied once, it can certainly be done again and again. The same Darwinian imperatives that today urge you to have children and spread your genes might drive you to spawn as many copies of yourself as you can afford to support. Luckily, you won't have to put your synthetic clones through school and university, buy their clothing, feed their growing brains and muscles. From day one they are as adept as you, and can earn their own keep.

Competing with yourself

If you are certificated as a lawyer or an electrical engineer, your copies will share your credentials, assuming the certification authorities manage to get their heads around the idea of cloned qualifications.

It is perfectly fair. Copies know exactly what the original knows, possess the same competences (and failings). You can jointly open a firm composed of yourselves, assuming you can stand their company, as partners. The new uploads can amortize their replication costs by working for half-wages until the bill is paid. Of course, it might now be harder to find work, because you have just flooded your local environment with competitors no less skilled than yourself. Will your surrogates be loyal to each other, or fight like competitive brothers and sisters?

Married to your computer

Uploading is going to force radical changes in the way people live together.

Such convulsive social change is already happening at a pace that might have seemed preposterous only decades ago. For example, in December 1996, a judge in the state of Hawaii ruled that same-sex marriages are legal. After two decades in which gay and lesbian activists had campaigned strenuously for equal rights under the law, this was none too soon, but it would have seemed utterly impossible as a political prospect even in the supposedly liberated sixties. Things will change even faster after upload. With machine-uploaded identities proliferating, Hanson notes, we might see multiple "team families" made up of the good old gang at the office, people who work or play well together. So the democratic principle of "one citizen, one vote" will become curiously blurry.

Uploads by the billion

Perhaps the most unexpected element in Hanson's scenario is "the potential for a huge [upload] population explosion." A flood of uploads might outrun the rate of growth of their joint incomes. A conventional evolutionary analysis suggests that populations tend to grow as fast as they can, consistent with available resources. More exactly, in Hanson's words, "Darwinian arguments suggest that if values can be inherited, then after enough generations the values in a species should evolve to favor the maximum sustainable population for any given technology, and the maximum sustainable growth rate as technology improves." How would that apply to uploads purchasing or funding copies of themselves? It's hard to say, since we don't know how expensive the process would be.

If it depends on nano minting, and if that is cheap, I can see nothing to stop an explosive proliferation of machines filled with the copied minds of those who love mirrors and others whose skills make them highly sought-after.

Hanson makes no such assumption of inexpensive copying. "The upload population could grow as fast as factories could generate new upload brains and bodies, if funds could be found to pay these factories." That conservative proviso—I can imagine skeptical readers smiling, but in the context of an imminent Spike it *is* conservative—simplifies the analysis, making it analogous to a property settlement after a no-fault divorce, or a couple's decision to have a new child.

In fact there's less anguish than you'd find in a divorce settlement, because both (identical) parties are certain to agree—at least, at first. Consider: if you turn out to be the version who has to forfeit the job and existing family, you might sour on the deal after a few days of loneliness. On the other hand, you'd readily admit that your very existence is the result of a choice you can remember making, even if you were not the individual who made it.

Upload families, Hanson observes, might jointly choose to restrict the numbers of additional copies they will permit. If hardware storage capacity runs into limits, specialized backups might be trained in narrow skills and then held in reserve for occasions when those skills are needed by the group.[12]

Invasion of the mind snatchers

How is one to guard against pirate copies? This sounds ludicrous, but taken seriously it's truly nightmarish. "An upload who loses even one copy to pirates might end up with millions of illicit copies tortured into working as slaves in various hidden corners. To prevent such a fate, uploads may be somewhat paranoid about security."

How do you block rapid illicit copying? For a start, you might have to abandon the blissful pleasures of electronically "teleporting" yourself from one platform to another, unless you have superb privacy encryption. It's one thing to be phone-bugged by government or unscrupulous private interests. How much more galling to awaken in the middle of faxing yourself to Africa and find that you have been mind-snatched into the software dens of some squalid upload sweatshop.

Hanson pursues his Darwinian analysis of the evolution of values under such unprecedented conditions, and tracks the rate of uploading through to a final equilibrium close to total poverty. "As wages dropped, upload population growth would be highly selective, selecting capable people willing to work for low wages, who value life even when life is hard. Soon the dominant upload values would be those of the few initial uploads with the most extreme values, willing to work for the lowest wages."

Upload cyberspace is a place where hard frontier ethics appear to merge with Marxian predictions of the gloom of unchecked capitalist contest. Hanson denies that this will lead to a glut of uploads willing to work for practically nothing, but does see his value evolution tending to "eliminate consumption of 'frills' which don't proportionally contribute to maximum long-term productivity." It sounds like a bleak prospect, with uploads vastly outnumbering ur-humans.

Some will see the way out of this impasse as a literal bursting free of the frontier, with uploads moving into ever-smaller and more powerful hardware substrates (for the Spike will still be accelerating), or outward from the earth into space. In principle, an upload has the happy option of controlling the clock speed of his or her mind. An interstellar journey of fifty or one hundred years at a modest tenth of the speed of light, too slow to yield useful relativistic time dilatation, could be made to pass in a subjective day or two, by running your simulated brain very slowly. Meanwhile, alas, your commercial and intellectual rivals might take the opposite tack, ramping up their processors to a thousand or a million times faster than an old-fashioned protein human brain.

The end of the upload era would come with the arrival of equally cheap human-level or superhuman artificial intelligence. Perhaps, Hanson speculates, we shall by then also know how to split off "partials" to roam through the greater mental ecology, doing our bidding before returning to merge with the stem mind.

In the meantime, though, we'd face some interesting political choices. While uploads might generally be highly competitive, if only because they can run faster and enhance their brain power with optional plug-ins, this doesn't mean that old-fashioned flesh types will be driven into poverty. If all else fails, they can put their savings into loans to newly minted uploads and thus share in the bounty of upload wealth.

Beat the rush—upload now

Will everyone have the chance to make a living as an upload? Perhaps not. Here's a rather disquieting possibility: "The first few uploads might have a strong advantage over late-comers; early uploads may have lots more experience, lower costs, and may be a proven commodity relative to new uploads. Billions of copies of the first few dozen uploads might then fill almost all the labor niches." This is the well-known *early-adopter advantage*.

One class of early adopter is sure to be the military, for unlike fundamentalists who merely *believe* they will be with their Lord if they die a martyr's death, backed-up soldiers will *know* that a copy will be minted if they perish in battle. Not that this means a lot if, like me, you can't accept that a perfect copy is truly continuous with yourself. Generals will be pleased, of course, for none of that expensive training will be lost with one injudicious fragment of shrapnel or laser beam flash thrust into a soldier's brain.

How might ordinary people stop these upload supermen from taking over the world? In a way, that strikes me as a little like asking how ordinary people can prevent the rich and well educated from taking over the world. They can't, and by and large they don't want to. The ideal in a modern society is to school your kids so that they *are* the well educated, and to put away as much as you can spare so that they will have a good start and you'll have the security you need in old age. Why would this change just because your offspring are uploads of yourself or your significant other(s)?

Here's a sensible recommendation, just in case we allow our gaze to slip from that logic: integrating uploaded people into society is surely the best way to keep the peace. Allow them to live and work among others in the community, handling their financial affairs through the same institutions used by unmodified humans. Exiling uploads into space or the oceans, as Hans Moravec has suggested, might be ill-advised. Actually, this notion has its merits, but I can see why Hanson is worried by it. Certainly anything that smacked of racism or slavery must be nipped in the bud. "Imposing especially heavy upload taxes, or treating uploads as property, as just software someone owns or as non-human slaves like dogs, might be especially unwise."

Living as Tinkerbell

Perhaps the strangest and most enchanting element in Hanson's appraisal comes when he blends three near-Spike concepts: first, upload techniques; second, nano minting's power to build small, dense gadgets; third, *tele-presence* (the kind of thing you can get right now by donning headset and datagloves and moving your point of view and action into a distant robot manipulator). His logic is a pleasing mix of the coolly rational and the madly runcible.

Very fast uploads will operate at subjective rates very much more rapid than protein brains. So rather than slowing them down arbitrarily, it might be best to kill two birds with one stone: load your point of awareness back down into a humanoid body built to the right scale. That might be just seven millimeters tall, arms and legs blurring at 260 times the rate of yours or mine, and running on 16 watts of power.[13] "Such uploads would glow like Tinkerbell in air, or might live underwater to keep cool."

Naturally, it might feel a little disconcerting to be so small, even if you run rings around the lumbering Big People. So the upload Lilliputians would tend to live in cities of their own devising and scale, perhaps attended by genetically engineered or synthesized miniature flora and fauna, and an artificial sun rising and setting 1800 times a week . . .

These pixies will peep and squeak at each other like bats, but their eyesight will be the poorer for it, partly due to the diminished size of their eyes and partly because the available light frequencies will seem to stretch due to the rapidity of their visual processes. On the other hand, they will be able to fall great distances without hurting themselves. The rules for the Olympics will have to be rewritten.

Life as a large bush

Even stranger destinies have been imagined—forecast is perhaps a juster word—by roboticist Hans Moravec. In 1988, you'll recall, Moravec made the sensational claim that we could expect to see a humanlike supercomputer before 2010, and a cheap personal computer version by 2030 (although he's now revised these estimates to allow an extra decade). Moving on from that point, a place where many other bold thinkers prefer to stop, he described a possible fractal "bush" robot that might be the physical implementation of an AI, or serve as the improved body of a human upload.

Cast off the parochial limits of anthropomorphic design—two arms, two legs, a torso, a head, a mere two eyes, a paltry five fingers on each hand. This sort of rough-and-ready product of the chromosomes has its place, but it's nowhere near optimum. No, we can build a *better* body these days—and it looks like some kind of walking or gliding shrub, sprouting feathery antennae.

"If we start with a stem a meter long and ten centimeters in diameter and carry the branching to twenty levels, the bush will end in a trillion tiny 'leaves' . . . Unfolded, umbrellalike, it would spread into a disk a little under two meters in diameter, thick but sparse near the center, and thinner at the edge, with smaller gaps that taper off to micron spaces." If you enjoy the look, this could be *you* in the closing years as we roar up the Spike. Because, of course, such a distributed intelligence—every tiny module an independent computer, but combining at will into something much finer—might serve as the ideal vehicle for any upload who has overcome our current boring and provincial taste for the humanoid form.

Moravec makes a persuasive case for the superiority of his bush body. Infinitesimal vibrating manipulators at the tips of the trillion cilia would be nanoassemblers, able to build a new full-scale bush just as the bush itself can be compiled out of raw stocks by a colony of the smallest machines. Communicating by light or pressure waves, its tiny modules could disaggregate and buzz about as smaller specialized bushes, or perhaps swarm into something like JoSH's Utility Fog (but more hardy). Moravec's model is composed of parts smaller than mosquitoes, flying by beating their cilia against the sluggish air.

Arthur C. Clarke made the necessary point, back in 1962 in his superbly prescient book *Profiles of the Future,* that miniaturized people would not look like stumpy full-sized folks but like insects, since the laws of geometry would call for a modification in the basic human plan. Tiny humans would need the "unbelievably delicate legs"—and perhaps wings—of insects; their muscles would be correspondingly less powerful, and the mechanisms of respiration and temperature control would have to be retrofitted (p. 196). So the seven-millimeter Tinkerbells of Robin Hanson's scenario, discussed above, would look less like Julia Roberts in Spielberg's *Hook,* and more like *Chironomidae,* the midge.

Like Hanson's home-away-from-home for uploads, Moravec's bush robot is proposed as a host for transplanted minds. Mightn't we prefer to transmigrate into a genetically improved protein body, or simply to rejuvenate and upgrade our existing carcass? Not so. "A genetically engineered superhuman would be just a second-rate kind of robot," Moravec comments scathingly, one handicapped by its DNA program and the restricted bodybuilding options available to protein life. "Only in the eyes of human chauvinists would it have an advantage . . ." (*Mind Children,* p. 108).

Uploading bit by bit

How do you get from here to there, from your current fleshy humanity to a bushy incarnation? A Moravec upload begins with a robotic bush-robot hand literally entering your brain, mapped in advance by high-resolution magnetic resonance techniques.

Layer by layer, the scanned tissues are measured in every possible parameter, and those measures emulated in a waiting computer. Once the simulated brain cells produce output results identical to the real brain cells they mimic, manipulators "excise the cells in this superfluous tissue." Bit by bit (literally), your brain is echoed into the computer. At last, your original brain is altogether gone—but your mind remains as active as ever. Finally, "the computer simulation has been disconnected from the cable . . . and reconnected to a shiny new body of the style, color, and material of your choice. Your metamorphosis is complete."

Oh, *right*.

Believe it or not, that is where Moravec's projection stops. (Although he does go on to sketch several other transfer methods "for the squeamish.") My own next sentence—penciled into my treasured copy of *Mind Children* after that last, astonishingly carefree comment—is this: "You are plunged into grief and mourning."

How can Moravec, a brilliant and compassionate fellow, not understand the violent sorrow that must attend what he proposes? Leave aside that jejune "shiny new body"; any woman who has lost a cancerous breast to the surgeon's life, even though the operation saved her life, knows better than that. Any parent who has seen a crippled, retarded infant die, even though that death spared parents and child alike months or years of futile torment, recognizes how fatuous Moravec is being.

Or is my complaint nothing better than the error of carrying over into a new, hopeful experience the unrelated deficits of earlier sorrows only accidentally similar?

After all, the shiny new body in a choice of colors and textures might now house the terminally paralyzed mind of a Stephen Hawking, or a paraplegic teenager otherwise doomed to a life in hell by a moment's inattention at the wheel of the car. Who am I to wag a moralizing finger at them as they move from that tragic condition to a reborn condition as a superhuman?

I find nothing intrinsically abhorrent in changing your bodily address in this way (assuming we can get past the suspicion that uploading is death by another name, which Moravec's slow transfer seems to answer quite well). It is the psychological brutality of the description I find distressing, the apparent lack of empathy. You might wish to dispose of your brain and body, you might even place yourself in debt for several centuries to attain that condition—but surely the moment of transition will remain, for most human beings, an occasion of loss and sorrow.

Mourning passes, though. What would it be like, this reborn life as an upload?

Being an upload

Once we learned the new ensemble of sensory windows, the range of motion open to us, we would less *inhabit* our new locus of consciousness than *be it*.

Whether as a bush robot, a glowing Tinkerbell, or a constantly morphing virtual presence flicking from one computer address to another, we'd surely stay much as we are—at first. Humans love the hint of risk, the taste of danger and exciting novelty, but we also require stability. Life is the unfolding of narratives out of chaos, and those stories tend to follow a powerful logic, strongly conservative.

That's a curious truth about the way we recall and construct the world that can be readily put to the test, as was done by Sir Frederic C. Bartlett back in the 1950s.[14] If a group is told a fairly complicated tale involving a trip down a river, rich with incidents both ordinary and weird, and quizzed a month or a year later, what do people remember? Not the oddities, oddly enough. The ghosts tend to drop out, along with other narrative elements that don't fit with conventional expectations. "The more people tried to recall the story as it really was, the more they conventionalized it," Jeremy Campbell reports, "until it settled down into a more-or-less permanent form in their minds."

Our hunger for sameness and safety only grows sharper when we're put in danger, deluged by change. So to begin with, most uploads will surely build consensus realities that resemble the settings of soap operas, or comfortably mock-exotic *Star Trek*, but probably not fractured movies like *Last Year at Marienbad*, written by Alain Robbe-Grillet and directed by Alain Resnais. Uploaded children and teenagers might be more adventurous, embracing the fluid *Matrix* medium joyously.

In the long run, though, as Moravec argues,

> Concepts of life, death, and identity will lose their present meaning as your mental fragments and those of others are combined, shuffled, and recombined into temporary associations, sometimes larger, sometimes small, sometimes long isolated and highly individual, at other times ephemeral, mere ripples on the rapids of civilization's torrent of knowledge.

These would be colony minds, spreading through the cosmos and absorbing everything, swallowing raw matter and transubstantiating it into life, mind, godhead. Of a sort. This is a post-Spike future with a vengeance, and of no immediate concern to us (although the most ardent transhumanists sincerely expect to be there, perhaps in multiple copies).

Still, our curiosity cannot help prompting us with certain extreme questions, and even more extreme possible answers.

Spreading through the cosmos

What happens to life and uploaded mind as it moves into the cosmos, leaving planets and individual bodies behind?

We might create ever more superior habitations for our thoughts and feelings, however repugnant that might strike us now. Moravec observes that if we ever learn how to sustain structures in the collapsed surface of a neutron star, a place where, in effect, all the electrons have been forced by immense pressure inside the nuclei to solidify as stupefyingly dense neutronium, our "clockspeed" would increase a millionfold, since our minds would be run on nuclear rather than chemical reactions. We might dub such a haven for mind a Neuron star. If life has evolved elsewhere in the galaxy, perhaps it has already colonized the available neutron stars, so that the spinning pulsars in the heavens might all be Neuron stars, thinking fleet and great and terrible thoughts.

Professor Gregory Benford, a sometime SDI ("Star Wars") expert who has served on NASA's Science Advisory Board and published over a hundred scientific papers, has projected a deep future for humans, machines, and more sublime minds. He suggests that the true high powers of the universe—the Phylum Beyond Knowing—might be memetic entities. *Memes* are Richard Dawkins's hypothetical ensembles of fads and habits and tropisms and concept-schemata that flit from mind to mind, rather in the way viruses infect bodies, replicating and mutating as genes do. In some sense our minds are rococo traceries of memes, or perhaps parliaments of contesting memes, each idea-module tussling with the rest for dominance. They may be expressed in action or simply in awareness. Memes sway our opinions, or even comprise them.

Benford's bold conjecture is that meme species, *kenes,* might somehow subsist at a higher level of complexity and authority than the relatively crude information systems native to brains. Kene mentalities would literally consume ideas for their sustenance, drawing energy from them as we draw strength from proteins and sugars stored up by the creatures we predate upon. These higher beings—or parasites, if you prefer—would be "datavores," eaters of processed information. And their relation to ontology, the deep reality underlying the phenomenal universe, might be at once unutterably mysterious and yet apparently magical, even godlike. They will not negotiate or dictate, Benford says. They will simply cause things to happen.[15] They will be the beings who follow humanity after the Spike.

7: The Interim Meaning of Life

I HAVE HAD IT. I HAVE HAD IT WITH CRACK HOUSES, DICTATORSHIPS, TORTURE CHAMBERS, DISEASE, OLD AGE, SPINAL PARALYSIS, AND WORLD HUNGER. I HAVE HAD IT WITH A DEATH RATE OF 150,000 SENTIENT BEINGS PER DAY. I HAVE HAD IT WITH THIS PLANET. I HAVE HAD IT WITH MORTALITY. *NONE* OF THIS IS NECESSARY. THE TIME HAS COME TO STOP TURNING AWAY FROM THE MUGGING ON THE CORNER, THE BEGGAR ON THE STREET. IT IS NO LONGER NECESSARY TO CLOSE OUR EYES, BLINKING AWAY THE TEARS, AND REPEAT THE MANTRA: "I CAN'T SOLVE ALL THE PROBLEMS OF THE WORLD." WE CAN. WE CAN END THIS.
—ELIEZER S. YUDKOWSKY, "STARING INTO THE SINGULARITY," 1999

How far will we be able to transform ourselves before, during, and after the Spike? And where will human societies end up, once it is upon us?

Already medical research provides heart transplants, drugs to enhance or control mood (however clumsily), dietary and advanced exercise regimes that allow us to morph our bodies, limited genetic engineering. As long ago as 1990 a four-year-old named Ashanti DeSilva, suffering from severe combined immunodeficiency caused by two copies of a defective gene, was given normal copies of the gene in a series of transfusions. "She has now been transformed from a quarantine little girl, who was always sick and left the house only to visit her doctor," reported one of her genetic therapists, W. French Anderson, five years later, "into a healthy, vibrant nine-year-old."[1] A further four years later, Ashanti was still well—but not strictly *cured*. More remarkable still, in the last week of 1999, British doctors announced that several infants with fatal Severe Combined Immuno-Deficiency

(SCID), a single-gene disorder preventing them from making T-lymphocytes (white blood "killer" immunity cells), are the first to be totally cured by gene therapy. A benign retrovirus carried a corrected gene into bone marrow harvested from the ill children, and when this was injected back their bodies gained the ability to express the normal form of the gene.[2] Meanwhile, French Anderson awaited permission from an advisory panel to insert corrective genes into fetuses with this disorder.[3] The early promise of gene therapy has not yet been fulfilled, then, due to the vigilance of the body's immune defenses. Worse, Jesse Gelsinger, 18, died in 1999 after a therapeutic attempt to infuse healthy genes into his damaged liver.

Further prospects along these lines, though, have the potential, via gene-vectored, enzyme or nano repair, first to repair defects such as the SCID gene, and ultimately to alter what it means to be human. We can expect that advanced technology will provide physical rejuvenation and perhaps immortality (except for death by accident or choice), increased intelligence and emotional depth and range, machine-enhanced memory, and the power to realize in virtual reality anything we are able to imagine.

Effectively eternal life

Every philosophy to date has been based on one of two postulates: either we all must die and vanish forever, or we all must die and pass onto some finer supernatural realm. The Spike, by contrast, will offer a genuine alternative, or an array of them: unlimited time on Earth and eventually, perhaps, among the stars, as physically youthful and fully mature transhumans. That utopian agenda might apply to *us*, personally, if we're still here as the Spike accelerates—or have providently arranged to have our corpses frozen in anticipation of post-Spike resuscitation. I have explored this prospect in detail in my book *The Last Mortal Generation,* which suggests that *we* might *be* that last generation doomed to die, or perhaps, if the research moves along fast enough, the very first generation free of the inevitability of aging painfully before, finally, hitting an unnegotiable expiry date.

In the meantime, we need to contrive the right mix of strategies and tactics to live in a world where technology is running faster and faster but its apotheosis seems always to hover out of reach.

The receding Spike

There's a perfectly sound reason why a technological singularity might remain on the horizon for some decades to come. It's not that the Spike, like some secular and sanitized version of the Rapture, is nothing more than a pious fiction or hope. Nor is it the case that the Vingean Singularity is a political figment, something we yearn to see but never will, like a secular version of "pie in the sky, by and by." When the Spike is well within range, perhaps in the 2030s, we will know it by the rush of its wind, the speed of change in the machineries of knowledge and power.

No, the point that tends to elude us, because we have had so little experience of it, is this: with exponential change, the biggest effects are crowded into the last few doubling periods.

Here's a homely analogy. The difference between strolling at one kilometer per hour and two kilometers per hour is not all that startling. And if you find yourself forced to wait four minutes for a bus, it doesn't kill you to wait eight. Really, what's another four minutes? But extend such a curve and eventually the doubling gap is between, say, traveling at 64 kilometers an hour, a respectable pace in a car, and a roaring jump up to 128 km/hr. In three more doubling steps you're traveling faster than sound.

So, while we've already started climbing the lower slope of the Spike, it's easy for us to fail to notice what's in prospect at the sharp end. Gregory S. Paul and Earl Cox make the point in two bludgeoning statements, both of them hard to grasp fully, both impossible to deny.[4] Here's the first: "We learned half of what we know about brains in the last decade as our ability to image brains in real time has improved keeping in step with the sophistication of brain scanning computers. For the sake of argument, let's assume we know just two percent of what there is to know about our brains right now. With the knowledge base doubling every ten years or so, we may know most of what can be known in about half a century."

If that estimate seems extraordinary (as indeed it is), consider this truly flabbergasting fact about computers: "We have already come a trillion times closer to human brain speed since the turn of the century, and the gap is now only about a thousandfold!"

Such numbers really can't be taken in, because their implica tions are so close to the invocation of magic. But the facts they encode imply, as Moravec saw years ago, that computers will attain human equivalence in just twenty more doublings. That's thirty years, if Moore's law keeps chugging away—or perhaps just twenty.

So we can't avoid the political consequences of this alarming insight. Those have already begun to bite, precisely because Moore's law is *not* sorcery but just a calibration of the accreted changes that technology layers down across the calendar, each change written in convulsively changed lives as well as ever more miniaturized chips.

What happens to people caught on such a runaway engine? Plenty. We need to look hard *right now* at the consequences of upcoming change.

Two implications of the Spike

First, the transition is already well begun, and needs to be addressed *without delay*.

The second implication, somewhat strangely, bears an apparently contradictory message: since most of the major changes will happen in the last few doublings, decades hence, we need to keep our feet firmly on the ground for the time being. There are choices to be made, political and personal decisions to be reasoned and argued through. Classic ideologies and the platitudes of the ages won't help us as we sway.

A world without jobs

Consider the following item of techno-utopianism, which you might suppose came from *Mondo 2000,* or maybe *Wired:* "There is only one condition in which we can imagine managers not needing subordinates, and masters not needing slaves. This would be if every machine could work by itself, at the word of command or by intelligent anticipation . . ."

Commonplace public relations from an information technology advertisement at the start of the third millennium? No, because there's gristle embedded at its center. We find ourselves unable to swallow the idea of a *world totally without work.* Still, we also find ourselves reluctant, I believe, to reject it as impossible fantasy. After all, a life free of toil is among civilization's oldest dreams. I mean "oldest" literally, because I cheated. Actually that passage was written by the philosopher Aristotle, in *The Politics,* over twenty-three centuries ago.

Granted, not even the spectacular synthesizing genius of Aristotle could foresee computer-controlled automation, let alone molecular nanotechnology. His remark, by a rather nasty irony, was a pragmatic defense of slavery in his own stratified society. We've seen marvels since then, but few greater than the general abolition of involuntary servitude. For the last quarter century we've hovered on the verge of a most remarkable achievement— the abolition of both poverty and the curse of distasteful toil.

The end of paid work

No more poverty? No more want and famine? Is it morally outrageous to make that claim while so many of the world's living human beings, half of them children, are at or below the brink of starvation?

Obviously enough, the abolition of toil and want is bound to come first to the advanced industrial and postindustrial countries. Yet despite a steady rise in the fortress mentality, in rich-world isolationism, ridding the wealthy nations of indigence and eventually of all arduous and repetitive work hardly implies any decrease in the aid they can provide to poorer countries. On the contrary: the very techniques of microprocessor control, fuzzy logic algorithms, expert systems, and sophisticated automation that give the first prospect plausibility also promise a general abundance that might alleviate the misery of less well positioned societies. AI and nano minting will make that promise even easier to meet.

Hoary traditions tell us that the sole cure for penury is work, and the harder the better. The acutely impoverished, unless they are the chance victims of flood or drought or war, are stereotyped as shiftless folk who cannot or will not work at an honest job (and are too incompetent to succeed even at a dishonest one). The obvious solution to the plight of those without money or property seemed to stare us in the face: discover or create work for them. Dig the fields, paint the barn, work the loom, run the messages, clean the chimneys—with your own body if need be. Get a job!

The age of downsizing and computers has turned that sanctimonious diagnosis on its head. As long ago as the 1960s, high point of the postwar boom days, a veritable army of the technologically redundant started to emerge. Many of those dismal tasks once so reliable in sustaining the undereducated poor, the various underclasses in First World nations, were being accomplished more cheaply and efficiently by machines. Within the last decade, high-level jobs have started vanishing as well, swallowed up by flexible computers, or shared out to people who would rather not do them. Typing pools were swapped for word processors on every desk, bank clerks replaced by automatic teller machines, architectural draftsmen and -women traded in for CAD programs.

Cyber Nation

In the fifties, the glossy magazines preached a coming age of "cybernation." It didn't happen, not quite, not then. Wage-winners doffed their felt hats, smiled at these futuristic fantasies, lit up their pipes, and turned to the black-and-white television in relief. The cyber word vanished amid giggles.

It has come back. Now it's here to stay.

Cybernated industries, it was once claimed, create as many new jobs as they make obsolescent. Probably true, although hard to credit. The stark and worrying reality behind *that* reality is simple: those new jobs cannot be filled by the displaced workers, or not all of them, nor can they be taken up by quite a large proportion of the young out-of-work. This tragedy has never been simply a matter of insufficient "retraining programs." Increasingly, the new jobs demand a level of education beyond the experience, and perhaps the competence, of many dispossessed industrial workers.

Meanwhile, near slavery returns in First World backyard sweatshops where entire families, often immigrants lacking in language and bargaining skills, struggle to compete on deregulated markets with imported goods from newly industrialized nations without much in the way of labor protection codes. None of this is stable, and it cannot go on much longer the way it is now.

Some segments of a temporary underclass, again usually immigrants, quickly burst free by community effort, family solidarity, and a tradition of respect for schooling. Education is plainly crucial in supplying work skills for those in extreme destitution— but, by the nature of the vicious circle of their lives, study is the last thing that most of their kids are eager to embrace. This is the age of guns and knives in the classroom, or at any rate in those classrooms where children most need to study almost anything but gang war and ethnic hatred.

Striking wealth inequalities exist everywhere. Professor Yew Kwang Ng, an economist at Monash University, reports that by a combined measure of income, natural resources, and leisure time, the richest countries in the world at the end of the century are Australia followed by Canada. A three-year study from the University of New South Wales found that, even so, Australia's wealth is massively unbalanced. More than a third is owned by one-twentieth of the families, and in total one-tenth of the nation's families possessed nearly half the country's assets.[5] Follow the money—it runs like sap up the family tree.

The significance is this: poverty in the First World, no less than the rest of the globe, is largely inherited. One generation passes it down like a tainted legacy to the next. Social security programs barely keep such subcommunities ticking over.

Are things getting better or worse?

This gloomy picture has been questioned. Karl Zinsmeister claimed in the *Wall Street Journal* that statistics paint a more cheerful portrait if rightly interpreted. During the last three decades or so, he stated, "hundreds of billions of dollars worth of health care, food stamps and housing transfers—which didn't exist in 1970" have been pumped into the U.S. economy, "unacknowledged in the national income statistics." Typical family size shrank by 12 percent in 23 years, making per capita incomes greater than simple comparisons might imply.[6]

An editorial in the same journal denied that the United States is a caste society locked in place by money.[7] Just 5 percent of those in the poorest fifth of income earners in 1975 were still there in 1991, which implied that mobility was vigorous. Nor had the definition of "poverty" remained static. In 1920, the lowest fifth of earners spent 70 percent of their income on such necessities as food, clothing, and shelter—and none of them (and none of the *richest* fifth, for that matter) had color television to console them in their misery. Thirty years later, the fraction of expenditure on the basics had fallen to rather more than half of the dollars earned by the poor, and by the 1990s it had dropped again to 45 percent.

A quite different interpretation of these data was offered by British socioeconomist Robert Theobald, to whose theories we shall return shortly. Theobald, who died at 71 of cancer in Spokane, Washington, in November 1999, was author and editor of many books and media presentations of a humanist, melioristic approach to human affairs (from 1959's *The Rich and the Poor* to 1992's *Turning the Century* and a four-part radio lecture series, *The Healing Century,* in 1998. His ideas had been in eclipse for a quarter century, but reemerged with a startling salience. His response was caustic, if light on detail: "There are lies, damn lies and statistics. It is quite clear that there are problems with statistics. But the errors go both ways. For example, as women move into the labor force, their time in the labor force is counted but the loss of work in the home is not." His rejoinder to the figures cited in the *Wall Street Journal:* "Look at the Genuine Progress Indicator [GPI] for another cut at how to measure the changes in the society."[8]

The GPI is a well-regarded instrument devised by a San Francisco Green policy research group called Redefining Progress. Its purpose is to remedy defects in the standard economic measuring tool, the Gross Domestic Product (GDP). That traditional estimate tallies an overall dollar value for the economy's marketed goods and services, but does less well in judging the economic welfare of the population.

The GPI, by contrast, incorporates such factors as unpaid housework and volunteer or charity work, while removing misleading components such as the costs of crime, pollution and resource depletion. "In other words," Chris Nelder has noted, "when the Exxon *Valdez* spilled oil into Prince William Sound, it showed up on the national books as a good thing. So does the Superfund cleanup debacle, Three Mile Island, crime, divorces, crass commercialism, excessive litigiousness, and most other kinds of social and environmental ills." You might wish to dispute some of these items—"crass" commercialism may be in the eye of the beholder, and divorce is not an unalloyed evil—but in general the distinction is stark. While GDP per head more than doubled in America between 1950 and the 1990s, GPI fell 45 percent since 1970.[9]

The rich, the poor, and the doomed

None of this, alas, should be news. It is instructive to see that some thirty years ago, economists Richard Cyert and William

Jacobs declared America's economy "technologically dynamic and distinctly uncongenial to those offering only basic labor and traditional skills—which is all the poor have to offer." This was long before the iMac, before the IBM personal computer and its vast illegitimate family of clones, before What You See Is What You Get, before object-oriented programming and mobile phones . . . The tens of millions then below the U.S. poverty line were already known to inhabit "a structure of numerous adverse and mutually reinforcing elements."

The key factor was that "the children of the poor are generally unprepared for entering school, and thus the children they raise are deficient in reading and verbal skills and in facility with mental concepts." As a direct result, the poorest 17 percent provided 40 percent of all school dropouts, and the poorest third some 70 percent of the total. Thus inadequately prepared, they were swept into a maelstrom of technological changes and demands that were, in total, the very world itself. The circle was impossibly vicious, and it has grown less benign as the skills needed to prosper have become more arcane, as crack neighborhoods turn poverty into hell.

For democracies, the bitterness is breathtaking. The old common presumption and boast of free-enterprise theorists, repeated anew in every confident declaration from "economic rationalists" and "level playing field" advocates, is that any parent might support a family in dignity if he or she were willing to take a job, even if the work proved to be dirty or hard. Today that is simply no longer certain. The life of the poor is a skein of part-time work, jostling schedules, social service entitlements that are repeatedly eroded or abolished, one or both parents on low wages at best and rarely at home for the children.

A social crisis has been upon the First World ever since the 1970s, much of it due to structural change, otherwise known as . . . (triumphant trumpet blast) . . . *the new technologies*. Most galling to many once comfortably off is the intrusion into their sedate security of forced redundancy packages, the scything of field upon field of middle managerial and skilled positions. The fate previously restricted to the "old poor," those who inherited their grief from their marginalized parents, has overwhelmed huge numbers of people with every reason to expect security of employment (and retirement). Ironically, this blight has afflicted even Japan, traditional home of the protected berth. The graying of populations makes this loss of job security ever more unbearable. "Japan will have a demographic profile similar to Florida's. By the year 2015, 1 in 4 Japanese will be 65 or older; by 2010, it will have only 2.5 workers for every retiree."[10]

At the same time, we need to remember that the superseded jobs are often precisely those involving the greatest degree of tedium, exhaustion, and lack of creativity. If people weren't losing their earning power as these jobs are swallowed up by machines, there'd be great rejoicing in the land. After all, in the abstract, the prospect of handing over to computers or manufacturing robots all those necessary tasks which are dirty, dangerous, tiring, and uninteresting would promise the human species a new era of freedom and personal growth.

Instead, as matters stand now, the dispossessed can look forward to nothing but misery, anxiety, and, by and large, marginal lives of boring pointlessness. Couch potatoes don't gobble Valium and Prozac because they're having a great time.

A guaranteed income?

One startling solution to these disasters is simple and feasible. It has been available for more than thirty years, but a moralizing prejudice stops us looking it in the eye. It demands no radical revolution, yet while answering the extremes of poverty that still afflict large pockets of the industrialized world it also opens the door into the coming dispensation, a world largely without jobs. And that's its drawback. It stands in profound opposition—or so at first it seems—to the traditional work ethic.

This is the suggestion, bluntly put: *society will pay everyone, as an inalienable right, a basic minimum dividend drawn from the productivity and wealth of the nation.*

Not just the ill. Not only the mad, nor the elderly struggling to bridge the gap between retirement benefits and costs. Everyone. No questions asked.

This is not a widely approved proposition. Although there is plentiful technical discussion in the economics literature, this option tends to be ignored or actively suppressed in serious policy debate.[11] Robert Theobald was an early proponent of this Guaranteed Income solution.[12] His mid-1960s insights on where Western society was headed, difficult to believe at the time, have turned out to be surprisingly accurate. Theobald and his colleagues foresaw an eerie future: an underclass excluded from jobs, a new educated class lurching from position to position with no assurance of secure income, and an elite working themselves into fabulously well paid exhaustion.

It was an astonishingly acute snapshot of the stressed First World at the start of the new millennium. Which gives hope that we might yet call the next thirty years with equal robustness, until the Spike intervenes. In 1992, happily unrepentant, Theobald wrote: "One recent poll shows that people would give up income for leisure if they had the choice. Personally, I find this development wonderful. I have long been in favor of full *unemployment,* believing that job structures were not the best way to get the work of the culture done" (my italics).

Beyond socialism and capitalism

Back in 1963, Theobald had published *Free Men and Free Markets.* "What practical steps need to be taken," he asked, "in order to reap the benefits of the scientific and technological revolution rather than its destructive growths?" His answer was shockingly offensive to conservatives of both left and right, and still is. He proposed "the establishment of new principles specifically designed to *break the link between jobs and income*" (my italics). Increasingly, the interests of the individual citizen, he reasoned, had been subordinated to those of the economy, filching freedom and dignity from a significant proportion of the population.

Forty years later, the frightful woes of the black and other marginalized "underclasses" in the United States, and the growing distress of those in a trap of welfare without the option of careers, prove how prescient he was.

The terms of Theobald's proposal were unequivocal:

> We need to adopt the concept of an absolute constitutional right to an income. This would guarantee to every citizen . . . the right to an income from the Federal Government sufficient to enable him to live in dignity. No Government agency, judicial body, or other organisation whatsoever should have the power to suspend or limit any payments assured by these guarantees.

Libertarians and extropians, those most open to the prospect of the Spike, will be horrified to read such a suggestion. This is heresy, they'll cry, for the State is vile. Government is incompetent, meddling, corrupt, taxation is theft, and handouts to the "work-shy"—extorted from hard workers—is a travesty of the proper order of things. Worse, it's counterproductive because it eats away the inner worth of the recipient, and tainted anyway, being the fruit of a poisoned tree.

One can share some sympathy for this distaste for bloated Big Government without immediately rejecting Theobald's suggestion. A guaranteed minimum income less resembles the demeaning receipt of an extorted, stigmatizing dole than the freedom to breathe the common air without paying for every breath. (Some libertarians, it's true, wish to privatize the air and oceans as well, in the free market belief that only thus will personal responsibility put an end to heedless pollution. Few seem persuaded by this extreme opinion.)[13]

How would Guaranteed Income work in practice?

Specifically, in the early 1960s Theobald examined the case of a family of four in which neither parent held a job. Drawing on official estimates of a "modest but adequate income" (since then wildly eroded by inflation), he suggested a basic annual income of $1000 per adult and $600 for each child. So the family would have earned, as a right, $3200 annually with no strings attached. Today the equivalent might be, say, eight times that much: a modest $25,000 in all.

So nearly everyone stops working at once, moves to California or some other sunny clime, and puts up their feet? Who pays? The few disciplined and self-respecting middle-class salary earners, gouged by Draconian taxes? Not at all. That Calvinist objection misses the point of Theobald's insight.

His suggested guaranteed income is enough to get by on, but hardly luxury. Note this key proviso: because basic income right has nothing to do with how you spend your time, you aren't *prohibited* from working for pay, or without pay, if you can find a job. To the contrary.

Isn't this, then, rampant socialism? Curiously, the notion was being promoted in the 1960s not by leftish radicals but by such saints of the free market as Milton Friedman, Nobel economics laureate in 1976. He favored direct payment to the poor as a way to slash the tangle of aid programs. A laissez-faire conservative and monetarist, Friedman argued that a guaranteed income would return to all individuals the freedom to seek their own economic advantage. Once that process was rolling, the magic of the market would supervene. Scrapping the complicated bureaucracy that stifled and humiliated the poor would lead to a revival of the nineteenth century's bustling free-enterprise drive.

Looking backward and forward

It was hardly a shockingly new idea, anyway. Edward Bellamy had raised the possibility in his 1888 utopian novel *Looking Backward*. Although the first industrial revolution had then not yet been thoroughly consolidated, Bellamy was able to foresee a world in which, due to mechanization, every citizen owned a "guarantee of abundant maintenance." It was a crucial leap of imagination to a society of *abundance* rather than *scarcity*. Despite recent slippage for some in the standard of living in the First World, *we are there already* if we choose to acknowledge the fact and act upon it.

It's a truism that consumer capitalism has to run hard and fast to sustain adequate demand for the relentless cascade of goods pouring from our factories. The pressure of this glut calls forth the huge, sophisticated machinery of the advertising industry, millions spent to persuade us to consume with lusty and extravagant zeal, and to go into debt in doing so.

And yet somehow the absurd capitalist machine *works,* balanced on its nose, pedaling as fast as it can, trawling up wonderful treasures and opportunities nobody has ever possessed in all the history of humankind . . . and bringing the Spike ever closer.

Shifting winds of opinion

Because of Friedman's support, Theobald found his early reception undergoing an acceleration rather reminiscent of the Spike, but with a nose dive at the end. Later, he wrote wryly in *The Guaranteed Income:*

> The bulk of the criticism in early 1963 evaluated *Free Men and Free Markets,* title not-withstanding, as an extreme left-wing text. By mid-1963, it was increasingly considered a modern re-statement of New Dealer philosophy. By early 1964 . . . the book was rather generally evaluated as 1964 liberalism. Finally, after 1964, the analysis and indeed the proposals are being characterised with increasing frequency as conservative and even reactionary.

Perhaps because of this protean character, the proposal remains provocative but untried. The core idea remains alive.

A simple version was devised by Cyert and Jacobs: a *negative* income tax, issued monthly. On all income from *other* sources, though, a family would be subject to high taxation, perhaps fifty percent. You could have a free lunch, so to speak, but you'd pay a little extra for dinner. Once the family's income reached twice the minimum rate, their guaranteed income would cut out, along with its confiscatory taxation level, and they would switch back to the ordinary graduated schedule of tax.

Is this impossibly expensive? No. The price of the program was estimated as about half the cost of a small war. In its absence, as we've seen, Americans are now in any case paying for a small war—against many of their own marginalized citizens.

As an inviolable constitutional entitlement, a guaranteed income would offer those who'd been made technologically redundant the chance to engage in dignified self-help. At present, it is difficult for jobless people to get a bank loan to start a small business, or buy and fix up their home, or pay for the protracted study and skill development that might lever them back into still-vital parts of the work force. "Further, and of potentially great significance," Cyert and Jacobs suggested, "a guaranteed minimum income would make it more feasible for private industry to sponsor long training programs, for it would no longer be necessary to pay substantial wages through the period of training and low productivity."

Consentives

Theobald was more optimistic still. Productive groups, which he termed "consentives," would come together on a voluntary basis, working *just because they wanted to*. Decades later, this is a pattern we recognize from garage bands, high-tech garage start-ups, and Internet special interest groups. All those, of course, tend to be funded by doting, well-paid parents. A different version is John S. Novak's notion, mentioned in chapter 4, of groups of comparatively young retired professionals with restless brains, around 2015 funding their own institute of yeoman tinkerers and innovative thinkers.

Consentives might produce goods that embody, both in themselves and in the workers' sense of creative satisfaction, the virtues of hand-crafted design in a machine world. Brilliant computer "shareware" and open source code does just that right now. Since wages and salaries would no longer be all-important, cost of purchasing such goods might be minimal, hardly greater than raw materials and transport—maybe comparable with computerized factory production.

The arrival of nano minting, naturally enough, will make even these fond hopes passé, but cheap minting will not rise above the horizon for at least another decade or two. We need drastic cures for the social problems that exist now and will worsen in the decades before minting makes them disappear for good.

After the work ethic, what?

National investment in a guaranteed income scheme might prove much more than a costly exercise in humane conscience-salving, and perhaps preferable to continued post-Cold War military stockpiling and planned obsolescence. Still, many will reject any redistribution of wealth, beyond a pittance paid for in humiliation. Poverty, it's supposed, is a character defect. If that claim was ever true, in the era of structural unemployment it is no better than cant.

The work ethic (better, the *job* ethic) cannot survive long while a culture of machine abundance disintegrates mores—themselves only, at most, a few thousand years old—of austerity and fanatical toil. It's undeniable, though, that a guaranteed income would run straight into the hostile defensiveness of those committed to the work ethic. Societies remain stratified, after the fashion of a scarcity economy, according to the jobs their members hold. Since the policy of full paid employment is doomed, we must do an about-face and perceive the merits of unemployment.

Fear of being out of a job has two roots that need no longer be axiomatic. One is the loss of adequate income. The other is loss of meaningful activity. Without the framework of discipline and satisfaction that brains and hands obtain from meaningful work, people start running amok with boredom, diverting themselves with the ancient, arbitrary, and zestful customs of tribal hierarchy and conflict. Force a generation of kids out of the loop, and expect them to trash your Porsche, murder each other, and burn out their furious grief with neurotoxins. This much is apparent to commentators from both the current left and right.

If Theobald was a thinker of the left, the late Australian political analyst and activist B. A. Santamaria was an old-style right-wing anticommunist one would not expect to advance such recommendations. Yet at the close of the twentieth century, Santamaria railed against the Organization for Economic Cooperation and Development, which advises that the nations it represents ought to lower tariffs in the name of free trade, cutting social services to the bone.

Citing *The Economist*, the journal articulating British capitalist opinion, Santamaria noted its recognition that "whole communities are caught in a deadly vise . . . work and marriage have declined together . . . these two, since time immemorial, have been the twin responsibilities that have persuaded men to stay with women and children, obey the law, and behave as social animals." That link broken, some men "become loose molecules: uneducated, unskilled, unmarried and unemployed." So mere subsistence is not the point. Welfare, for traditional conservatives like Santamaria, supports the body but not the soul. "What is needed is *not* welfare, but work."

Work without toil or jobs

Theobald would not have disagreed with the substance of this analysis, although his hopes lay in a renewal of community and purposeful activity, with the physical and psychic backup of a guarantee of essential security. An emphasis on the value of affectionate parenting—understood as important work worthy of pay—will continue to develop, he argued, since "the skills gained in parenting are directly relevant to the needed work" of the medium-term future, requiring subtle interpersonal skills. In a world mutating every more swiftly under the impact of high technology, detestable toil will be very hard to find, or even invent.

Santamaria's advice harked back to post-Depression solutions: "not merely make-work [but] municipal, shire, regional and national infrastructure projects," as modest as local roadworks, as large as grand national waterworks. Who was to pay for this program, and how (the same question posed about any guaranteed income scheme)? Why, the State, the collectivity of citizens acting through their elected representatives, and through the credit-creation potential of the central bank "at the minimal rates of interest which, by cheapening the cost of capital, have made possible the enormous industrial development of Japan."[14]

The invalidating assumption, of course, is that no more leaps in technological prowess are in the offing. Men (but not women, who apparently should stay at home happily with the kids) will drive the tractors that build the roads upon which the cars and trucks will thunder into the polluted distance. That prospect might just be plausible, if unacceptable, for another decade or two. Beyond that point, there'll be little authentic toil of any kind for humans to endure, however cowing or civilizing such activities might be deemed.

The dividend of history

The run up to the Spike will deepen today's dreadful problems, but also ease their solution, if we keep our nerve and use our brains.

A corporation that downsizes its work-force today, in favor of robots, is surviving as a beneficiary of the human investment of the past. Its current productivity, after all, is the outcome of every erg of accumulated human effort that went into creating the economy and technological culture that made those robots possible.

So let's not look at a guaranteed income as a "natural right," like the supposed innate rights to freedom of speech and liberty. Rather, it is an inheritance, something *owed* to all the children of a society whose ancestors for generations have together built, and purchased through the work of their minds and hands, the resource base sustaining today's cornucopia.

Committed to the present structure of society—a prejudice doomed in any case, as we have seen, by the oncoming Spike—some people believe that most of their fellow humans won't work without an external goad. Others agree that incentives are required, but hold that these can spring from within, and need not depend upon the threat of hunger and destitution. In the short run, this debate can be avoided, for a guaranteed income would abolish severe poverty more effectively than current schemes that tend to act as a disincentive to taking up part-time work.

And while a guaranteed wage would ensure you the bare necessities, your craving for luxuries and a higher standard of living would hardly disappear. Few of us now are content merely to earn a subsistence income. Those who are—the "shiftless scoundrels" who live their rudimentary life of ease on the dole in sunny climes—will continue to do so, but without costing society the extra burden of trying to hunt them down and punish them for refusing to take the jobs that don't exist.

What we must hope, as the juggernaut of technological change rolls onward, is that material incentives alone really are *not* more important than other driving forces of the human spirit. The artist and the scientist are two celebrated instances of a deep human hunger and enthusiasm for creative activity, the kind that draws people together while being intensely rewarding to the individual. It's very likely, no doubt, that some arduous jobs traditionally the domain of depressed and excluded groups would find no takers if survival were no longer at stake. In wealthy nations like the United States, bordering less well off countries, a constant influx of illegal migrants testifies to the reluctance of citizens with high expectations to do the sweaty work.

Return to the village

The impulse to automate such unpleasant activities, or design around them, is thus intensified. If nobody can be found to take the nasty jobs, the proportion of such unrewarding tasks delegated to machines will grow (thankfully) until people are free of them for good. And in many cases, we can expect that nano minting and AI systems will simply wipe many of the worst kinds of odious toil off the agenda forever.

Hans Moravec puts it succinctly: "In the short run this threatens unemployment and panicked scrambles for new ways to earn a living. In the medium run, it is a wonderful opportunity to recapture the comfortable pace of a tribal village while retaining benefits of technological evolution. In the long run, it marks the end of the dominance of biological humans, and the beginning of the age of robots"(*Robot,* p. 131).

The anthropologist Conrad Arensberg claimed that every successful adaptive innovation, social or biological, always has a further effect than the immediate and conservative: "the opening of a vast new door, a splendid serendipity." It is impertinent and finally futile to try to anticipate serendipity, but it seems fair to assume that the adoption of a general right to a share in the inherited productivity of the human race will be liberating, more often than not, to the spirit.

Prepaid lunch

Surprisingly, the laissez-faire writer Robert A. Heinlein placed that sentiment in the mouth of a utopian judge condemning a reckless rugged individualist: "From a social standpoint, your delusion makes you as mad as a March Hare." But it was Heinlein himself, coiner of the libertarian slogan TANSTAAFL ("There Ain't No Such Thing As A Free Lunch"), who, as narrator, offered the most stinging commentary:

> The steel tortoise gave MacKinnon a feeling of Crusoe-like independence. It did not occur to him his chattel was the end product of the cumulative effort and intelligent cooperation of hundreds of thousands of men, living and dead.

Granted this perspective, how should we arrange our affairs during the disruptive decades ahead? The ways in which science and technology were (ab)used in the last century powered a catastrophic erosion of biodiversity, and in doing so damaged human communities as well. The global economy is still felling rain forests, polluting the skies, and scything through the diversity of species. In the long run, in a world of televised fantasy living and virtual reality simulation—let alone the Spike's machine intelligence and nanotechnology—such traditional methods will be altogether unstable. Should we put an end, for the moment, to conspicuous consumption? Yet it is exactly the economy of relentless and often trivialized consumption that drives the technologies that will peak in the cornucopia of the Spike.

The big picture

Global problems, such as the world environmental crisis, require the attention of global thinkers, as well as more precisely focused action by plenty of people at every available local level. But global thinking is in bad odor, for two quite different reasons.

In the sciences, as well as economics, the well-attested secret of success is specialization, expert competence applied with ferocious intensity to a well-defined and bounded problem. In the humanities, postmodern and poststructural analyses announce the extinction of grand narratives. Every knowledge, we're told, is situated, relative, political at the core. In neither realm, then, is there any place for global thinkers intent on overarching perspectives.

Jaron Lanier, sometimes credited with coining the term "virtual reality," has suggested a division in philosophies of change—approaches to the oncoming Spike, one might say—between Extropians and Stewards.

A Steward is somebody who wants to manage the world as a precious resource, and an Extropian is someone who wants to let some big, impartial evolution-like process run wild with it. Extropians differ about which process this should be, though it certainly can be the more traditional libertarian capitalism combined with the self-propelled on-slaught of new technologies. Extropians don't worry about natural resources running out, or about poverty, or any of the other problems that frighten Stewards, because they are convinced that new technologies will solve the problems if we just give capitalism and science an unfettered chance. Stewards speak a language of what's already here, like hu-man beings and rocks, while Extropians believe that every-thing here is going to be replaced by new, evolving things anyway.[15]

Lanier, curiously enough for such a whiz kid, tends to side with the Stewards on any "life-and-death issue, like global warming or nuclear disarmament . . . We should attempt to regulate and con-trol the hell out of everybody involved. But when there's room to experiment, as in almost all cultural matters . . . I'll risk Extro-pianism now and then and brace myself for the wild ride."

The politics of the Spike

We cannot ignore the large-scale political coloration of these dis-ruptive innovations. As self-bootstrapping AI and minting begin to affect our experience, or perhaps the experience of our children and grandchildren, it is certain that the international order built up over the last six hundred years will crack and crumble.

Not that it has been terrifically stable until now—wracked, rather, by regular convulsions of hostilities and outright wars, ceaseless trade conflicts, jostlings for spheres of influence, tech-nological spurts that have advanced the ease and personal power of hundreds of millions and helped murder as many millions in unprecedentedly horrible ways. Is there any way to map out in advance the impact of the pre-Spike acceleration?

Perhaps it is not altogether impossible. In a fascinating paper,

Thomas McCarthy has tried to do just that, testing the boundaries between today's polities and the emerging molecular nanotechnologies (MNT).[16] McCarthy is active in the nano field as president and CIO of Ntech Corporation, the first publicly traded nanotechnology development company. "The companies that make up Ntech are shipping real products today," McCarthy states, "but are focused on the long-term development of nanotechnology." Before taking up his current post at Ntech, McCarthy was coordinator of international relations for the Nagano Prefectural Government in Japan. If his still-unfinished political analysis leaves out of account the impact of AI and genetic engineering, which really ought to be factored in to complete the story, perhaps that is just as well. We need to impose some ruthless limits on these scenarios, or risk being driven into idle fantasies of omnipotence.

"On a global level," McCarthy observes at the outset,

> MNT poses a serious challenge to decision-makers and ordinary citizens alike. On the one hand are the enormous benefits that the technology may produce, from personal and national wealth that nearly exceed our ability to imagine, to possibly final cures to the aging and diseases that have plagued us for our entire history. On the other hand are the dangers, hard to visualize at this point, but all stemming from the radical power of this new tool. Whether superweapons of unspeakable destructiveness or home-made attempts at genocide, at least some of the potential products of an advanced molecular manufacturing ability give us pause, making us wonder if the heralded "Age of Nanotechnology" will be nothing more than a finely engineered disaster.

At best, minting might remove many of the age-old causes of conflict. At worst, it will destabilize every traditional balance of trade and terror, the carrots and sticks that have tended to hold nonstop aggression at bay.

War and memes

Wars have not always been over anything so negotiable as the ownership of raw resources and lucrative tracts of land or sea. Often conflict is memetic, driven by ideology, philosophy, and theology, sometimes by abstractions so remote from reality that we can only shake our heads at the insanity of the combatants. Did it *really* matter so much whether Jesus was of *one* substance with the Father, or only of *similar* substance? Many hundreds of years after the Arian heresies and the terrible slaughters they occasioned, few Christians find any meaning at all in this archaic philosophical analysis. Yet it became the cause of hideous bloodshed.

No doubt earthier issues were also at stake, matters of raw power and domination. Still, the ideological differences were not a mere mask or blind for the *realpolitik*. People believed to within an inch of their lives, as they do today in Iran and Iraq and Serbia and Ireland and Bible Belts in various otherwise enlightened nations. In their ferocious, embattled belief they fought and defended themselves to the death of their children, and still do.

Nanotechnology cannot have any impact on that kind of memetic contention, except perhaps by easing the psychological pressures of want and fear that help fuel such thought-viruses. It can, however, directly influence other factors: available resources, manufacturing capability. McCarthy shows, in a series of well-argued and alarming moves, that minting is as likely to unsettle the next few decades as to bring the world into peace and plenty.

Weapons supply made simple

To begin with, warrior culture is contained by the speed with which weapons can be acquired or produced. Until now, the only way to ready a modern nation for war was to stockpile arms, and that has been a treasury-exhausting business with long lead times. Bombers are decades in the planning, obsolescent before they roll off the production line. Missiles, like domestic air services, are controlled by computing systems generations out of date. Minting alters all that.

Mature nanotech will enable belligerents to mint tanks and fighter jets and missiles from algorithms and raw materials in a matter of days. But such weapons themselves will be antiquated. The same invisibly small machines that run the matter compilers are themselves the ultimate terror weapons, when programmed to deal death.

Loose selectively murderous nanites at your enemy's borders (*war goo!*) and within days or hours they will have rushed into the heartland, poisoning or killing or disassembling everything in their path, a locust plague on the twin scales of molecules and hurricanes. And the early chances of detection of these insidious foes is minimal.

True, the equivalent of antibodies against the menace of nano weaponry will surely be developed—*active shields,* Drexler calls them, a form of blue goo—and their minted production automated. But the speed of infection by a nano blitzkrieg might melt everything in its path. More to the point, as McCarthy notes, the target of aggression will once more invoke the heinous doctrine of Mutual Assured Destruction, only recently put aside between the great powers.

Who is the target for nano weapons?

If your enemy's resource base is just the common elements freely, abundantly available in the soil and the air, if their factories are hundreds of thousands of small goop pools or domestic matter compilers, what target is there for you to bomb, to interdict, to raze? The target of choice becomes, once more and horribly, the citizenry themselves. Us.

It is the nightmare once symbolized by the neutron bomb—a weapon that strikes preferentially at living organisms and slays them—only this time converting them into neat piles of raw materials, while sparing *everything* that's worth keeping: the farms and livestock, the housing, the art on the walls, the humming laboratories . . .

Deterrence, McCarthy reminds us, *works*. Knowing the enemy's logistics means that you can estimate your own safety. That is why the Great Game of espionage has always been as much an aid to peace as a hindrance. That is why mutual inspection of hyperlethal arms, and satellite surveillance, is accepted by great nations and small alike.

Nanowarfare, especially if it's amplified by machine consciousness, snatches visibility away for good. And when you can't see what the enemy is up to, your justified suspicion readily festers into irrational paranoia. There is the temptation to *show them what you've got,* to stage small atrocities in a highly public way under some convenient *casus belli,* and trot out the new horror weapons. It is a temptation we have seen nations submit to repeatedly, even if nothing is destroyed but an evaporated South Pacific island and the local children downwind of the "test" fallout plumes . . .

Superpower or diversity?

To protect ourselves in this next unstable period, while people get used to the possibilities of genuine abundance and decentralized production of goods and power, nations need to learn to find (forgive the sermon) unity in diversity. The sole alternative is for one global superpower to impose its stamp upon the rest, quite the opposite of what we see occurring today.

Yes, the world is swaddled in thickening electronic links, of Net-mediated finance and meme-rich entertainment vectors and the sheer inundation of information, mostly in English. Yet its traditional political unities are fracturing, fraying, breaking apart once more into tribalisms. Communities are still based on crude signs of identity: common faiths, skin color, language, region. It is a dynamic I find altogether predictable, and I expect it to continue for the next fifty years—the very interval during which we expect to see the Spike accelerate into the unknowable.

We face a dilemma, then. Peace and truce require mutual dependency, the sort of thing encouraged by markets. Once, each village made what it needed, and traded only for ritual objects and curious luxuries. Machine civilization showed how specialization might enhance our ability to make many more things and exchange them across vast distances. It showed how we might seek out rare stocks of resources and build ever more consumables and durables with them. Minting transcends that market pattern, for good and ill.

The death of the market

If your town can make most of what it needs in the local goop pool (and if, a few years later, you can do most of it at home), and if power for your machines derives from efficient, cheap nanominted solar cells or genetically engineered fuel crops, the market collapses—or at least it contracts savagely into specialty items.

Your community moves into what Thomas McCarthy calls Autarky, a condition of local self-reliance. Minting will throttle the economic need for large states, and provide fertile ground for a swarm of new, small states. What we'll find, if this works out, is an anarchic mix of *designer communities*.

> The concept of designer communities should not be as surprising as it might sound at first; after all, there is ample precedence for it, found in the on-line world of bulletin boards, chat rooms and MUDs [multiple-user domains]. For anyone unfamiliar with these terms, a brief, yet comprehensive, way of explaining them is hard to produce; they are best experienced first hand. Nonetheless, the core concepts of the communities formed via the Internet and other electronic means of communication and their relevance for designer communities can be captured in this phrase: they are *virtually costless* communities. In this key respect, they resemble the state of tomorrow.

Small democracies of this kind—or, indeed, small vicious armed enclaves of racists and religious extremists and who-knows-what-all—will crystallize out of the gradual disintegration of unwieldy, unnecessary giant states.

Those megastates and their military and police forces and the powerfully rich (legitimate or mafia) will not subside gently. Nonetheless, the sight of Russia's once-mighty armed forces scrambling for a place after the fall of the empire, soldiers selling blood to provide food for their children, suggests that power can slip away even from those armed with terrible weapons.

Leaving Earth behind

Ultimately, McCarthy suggests, a post-Spike civilization (although he does not put it that way) will drift into space. Not merely to colonize nearby inhospitable planets and moons but in a great teeming of "self-contained worlds, existing independently of a planetary surface." This apparently absurd prospect has been sketched in charming and substantial engineering detail by Marshall T. Savage and his associates, as we shall see in the next chapter, and there seems little doubt that a world with the limitless resources of AI-coupled nano minting could make the leap into solar space.

Why would anyone wish to do so? We love the earth, Gaia, our mother world. Would we leave her? Once more, McCarthy's cool analysis probes our prejudices. Political proximity of independent states, as the philosopher Immanuel Kant pointed out long ago, is itself a source of friction and distrust. In a minting world, there might be no particular brake upon human population growth, since industrial abuse of the environment will be a thing of the past, rendered profitless by the new technologies. But even with increasing genome control, it is unlikely that we will soon change ourselves as much as we change our circumstances.

So the Kantian antagonisms will persist. People will chafe as their neighbors rub blue woad into their bellies, or build telescopes and particle accelerators despite the express disapproval of the local tribal religious leaders, or vote for their leaders by secret ballot, or whatever other way of life happens to strike your particular affinity group as heinous and intolerable.

If it is technically feasible simply to up and leave, hurl away into homesteads in a great belt of lushly appointed nanobuilt biospheres in the same solar orbit as the Earth, many people will take that option.[17] And their post-Spike descendants will inherit not the earth, as in ancient biblical prophecy, but the universe.

Making the Spike happen

Perhaps the most strongly worded expression of hope in the redemptive powers of the Singularity is an extended essay, "Staring into the Singularity," by Eliezer S. Yudkowsky. He is a prodigy who at 11 gained an SAT score of 1410 (higher than most college entrance students manage) and a dazzling score of 1600 at 15. A little later he turned his undoubted abilities to the customary adolescent task of fixing the world.[18]

Yudkowsky regards himself, interestingly enough, as an evolutionary mistake, a kind of mutant genius. He believes his undoubted abilities derive from accidental specialization, the diversion or conscription of neural capacity from some standard mental fittings to enhance the modules that power his special gifts. That's not impossible, and his efforts in introspection, conscientiously framed by recent findings from cognitive science, offer some surprising insights into the path transhuman intelligence might take as it scales the Spike.

No doubt the most disturbing aspect of Yudkowsky's crusade is his suggestion that we might bring the Singularity nearer, faster, by deliberately manipulating the brains of children to enhance specific gifts well suited, as he supposes, to solving the outstanding obstacles to AI and other Spike-conducive requisites.

Yudkowsky calls such damaged but gifted people "Algernons": "Any human who, via artificial or natural means, has some type of mental enhancement which carries a price."[19] He speculates horrifically that Algernons might be mass produced, "starting a year before puberty," by locating a specific ability in the brain (or rather locating, in the more modular limbic region, an emotion-module evoking that ability). Neurosurgical stimulation would then switch the ability on, ensuring that it was constantly active. Suppose a small lesion burned into your daughter's brain could turn her into a world-class pianist or gymnast. Would anyone do this to their child? It's worth recalling that parents do quite frequently design obsessive and restrictive lives for their children, organized around just such ambitions—musicians, swimming stars. It's also worth recalling that this behavior is usually regarded with contempt by other parents, however much they envy the eventual success stories and media glitz.

Side effects, Yudkowsky admits, would create cognitive blind spots and alter the normal development of the mind, but one gains the impression that such ethical atrocities pale before the need to get the Spike's ameliorative wonders here as swiftly as possible. In a more recent essay, "The Ethics of Cognitive Engineering," he claims his own "Algernic" status gives him the right to make the proposal. But should he ever come to regard Algernization as evil, he declares, he would abandon his proposal—even if it meant the loss of his beloved Singularity.[20]

That is a very large concession, for he sees the creation of the Singularity, or perhaps our quest for its realization, as a kind of ultimate moral urgency:

> Our fellow humans are screaming in pain, our planet will probably be scorched to a cinder or converted into goo, we don't know what the hell is going on, and the Singularity will solve these problems. I declare *reaching the Singularity as fast as possible* to be the Interim Meaning of Life, the temporary definition of Good, and the foundation until further notice of my ethical system.

Well, of course, one smiles, recalling in flinching memory the exaggerated postures of adolescence and idealistic, impatient young adulthood. Still, Yudkowsky's case has its merits. And in the midst of his tussles with the sublime prospects of a Spiking

world, he manages to blend old-fashioned drum-beating fervor
with ornery common sense:

> Right now, every human being on this planet who has heard
> of the Singularity has exactly one legitimate concern: **How
> do we get there as fast as possible?** What happens after-
> ward is *not our problem* and I *deplore* those gosh-wow, un-
> imaginative, so-cloying-they-make-you-throw-up, and just
> plain *boring* and unimaginative pictures of a future with
> unlimited resources and completely unaltered mortals.
> **Leave the problems of transhumanity to the transhumans.**
> Our chances of getting anything right are the same as a fish
> designing a working airplane out of algae and pebbles. **Our
> sole responsibility is to produce something smarter than
> we are; any problems beyond that are not *ours* to solve.**

The emphases, designed for maximum impact on his Web site,
are his. Yet this is not quite the pie-in-the-sky, after-the-
revolution rhetoric the conventional and the timid will mistake
it for. Yudkowsky wants the Singularity, and he wants it *now!*—
so, for many people, the argument will be hard to take seriously.
That might well be due to our workaday dullness, our commonplace failures of imagination, our inadequate powers of feeling
and thought when faced with the genuinely new and different.

The Spike will happen because people make it happen, Yudkowsky asserts, not by accident or simple convergence but
through hard dedicated research in a dozen disciplines, through
investment and effort. Meanwhile, front-line researchers are
ground down by committee work, by the time-wasting need to
fill out endless forms and justify themselves to accountants. Perhaps the key to the Spike lurks in the research and development
progress of someone toiling at the lab bench even as we speak.
Here is Yudkowsky's desperate plaint:

> *Every hour* that person is delayed is another hour to the
> Singularity. Every hour, six thousand people die, and most
> of the survivors are unhappy for an hour. Perhaps we should

be doing something about this person's spending a fourth of his time and energy writing grant proposals.

Suppose he's right? In essence, it is, after all, the same moral claim that thousands of cancer researchers make upon us and our funds, as do molecular biochemists striving to conquer the AIDS virus, and even those suspended animation researchers who deplore the loss of millions thoughtlessly buried or burned after death, who might instead be cryopreserved for later revival and rejuvenation when science has *got* the job *done*.

Perhaps the most impressive aspect of Eliezer Yudkowsky's brilliant, youthful enthusiasm for the Singularity is his own creative outpourings in attempting to forward it. Not content with vigorous propaganda, he has begun the daunting task of creating a seed AI, one in which the design allows for—indeed, encourages—a self-enhancing program. From rudimentary beginnings it would revise itself upward to human-scale attainment (by 2010, or even 2005) and, he hopes, onward to superhuman levels of intelligence. This project has yielded an interesting, quirky set of algorithmic strictures and guidelines, sketches for elaborate detailed programming, titled "Coding a Transhuman AI." What's more, with the support of a youthful Atlanta millionaire, Brian Atkins, Yudkowsky has become Research Fellow at a nonprofit Singularity Institute with the single and singular goal of bringing about a Spike as soon as possible, preferably before the arrival— from military researchers, or existing commercial labs such as Zyvex—of lethal molecular nanotechnology. (Initially a nano enthusiast, today he is convinced that molecular nanotechnology in the hands of unaugmented humans is an almost certain recipe for global devastation, perhaps annihilation.) His elaborate strategy is dubbed "The Plan to Singularity," and it is conceivable that he and enthusiastic supporters might manage to do just that. Unless he is closed down, taken over, or the task simply proves too immense for any individual or small group undertaking, as seems altogether likely.[21]

Yudkowsky is aware how preposterous his expectations might sound. In the spirit of Drexler's own public caution, he advises supporters "not to go Utopian. Don't describe Life after Singularity in glowing terms. Don't describe it at all."

It's a curious paradox, because Yudkowsky is actually quite effective at conveying something of the nature of any true Spike.

What I find most compelling is his intuition of the Spike's genuine strangeness. "Looking at stories of instantly healing wounds, or any material object being instantly available, doesn't give you the sense of looking into the *future*," he has remarked caustically, citing popularizing attempts to project the terrific time we'll all enjoy once minting is here. "It gives you the sense that you're looking into an unimaginative person's childhood fantasy of omnipotence, and that predisposes you to treat nanotechnology in the same way. Worse, it attracts other people with unimaginative fantasies of omnipotence."

If all this is so—and contriving a truly plausible future is extraordinary difficult, precisely because the reality is bound to exceed expectations, and wildly—then why should we care about the Spike? Yes, it might be there, just over the horizon. It might roar into history by 2035, or even sooner. But by definition it will create a discontinuity with the past—and with much we now hold dear.

The Beyond

Do we care? If transcended minds are due to be born of some mix of computer design and coding protocols and brain research and molecular nanotechnology, what will they be to us, or we to them?

Yudkowsky borrows several terms from Vinge's visionary fiction to capture and resolve this quandary. *Powers*—entities native to the far side of the Spike, perhaps human-machine composites—will possess the ability to rewrite their own capacities, to bootstrap themselves and their technology in a rapid feedback loop.

People not yet in that lofty state will be *transhumans* dwelling in the Beyond, unable to reprogram themselves into Transcension, the wondrous condition attainable only by posthuman Powers. They will not have to stay there, though, with their noses pressed up against the windowpane.

Armed with molecular nanotechnology (at the very least), Yudkowsky argues, it would be trivial for these Transcendents to uplift their progenitors, uploading and upgrading the entire human race to Power status. The Spike, then, would hold out the hope of "endless growth for every human being—growth in mind, in intelligence, in strength of personality; life without bound, without end; experiencing everything we've dreamed of experiencing, *becoming* everything we've ever dreamed of *being;* not for a billion years, or ten-to-the-billionth years, but *forever* . . . or perhaps embarking together on some still greater adventure of which we cannot even conceive."[22]

Are there any limits to this headlong dash into sublimity? Yudkowsky is surprisingly cautious, facing the peeling edge of exponential and hyperbolic change-curves. These curves, as we've seen repeatedly, are nothing real. They are not pressures or forces or causes. The wild upward curve of the Spike is a mathematical construct, a map of where we've been and, if nothing steps in to change the equation, where we are headed.

But, of course, processes of doubling cannot continue forever. The key postulate of the Spike is that each doubling (looking at the matter conservatively) takes the same amount of time, so we swiftly run into Harry Stine's 1961 error of extrapolation: each person on the planet in possession of a star's energy resources, say, or craft traveling at light speed, and all of this due before 1982 . . . Yet an even more dramatic interpretation sees the doubling itself *speeding up,* once self-programming machine or uploads are involved. A jump that once took a year to accomplish is now made in six months, the next equivalent jump in three, and so on. This is no longer exponential growth, but *hyper* exponential.

Does this make sense? Certainly, says Yudkowsky. Powers will be entities in charge of their own cognitive architecture. Suppose the calculational power on the earth today is some 100,000,000,000,000,000 operations per second in each human brain, and there are some six billion of us. Multiply that together, and compare it with all the computer power in the world, and it's glaringly obvious that our collective brains currently outstrip our creations by an inconceivably great extent.

However— As teraflop, petaflop, and still faster machines become common, that unbalanced equation will start to shift away from human brains and toward computers, then to AIs and uploads. It has begun to do so already.

After all, a few centuries ago there were hardly any machine calculators in existence. At some point, if Moore's law retains its validity, machine calculations will equal those occurring inside human heads (somewhere between 2020 and 2040, maybe sooner). A little later, the balance will topple over. Very quickly, "*humans* become a vanishing quantity in the equation."

Limits to the Spike

Constraints on what can be achieved will remain, however, not easily to be breached.

Repeated doubling fails sooner or later, because even if the researchers are AIs inside the world's fastest computers, presumably they can't advance to the next generation's physical manufacturing technology in seconds. It takes an appreciable time to assemble materials (even with minting) and rebuild circuits. A superb AI can't bypass the fundamental laws and limitations of physics (whatever those turn out to be once we truly have a Theory of Everything).

For now, though, we hardly need concern ourselves with those ultimate boundaries, assuming they exist. The Spike will certainly continue clawing its way up the exponential slope for many doublings before the technologies involved run into absurdities and curve over their inflexion points into a steady state. (Actually, Yudkowsky argues that the relevant technologies will *not* slow before halting. Rather, he simply expects them to ram up against the ultimate limits and stay there, running flat out forever.)

Beyond the Spike's barrier

Yudkowsky paints an intriguing portrait of the mind of a Power, while insisting that whatever we can say about post-Singularity minds is inevitably a travesty, a painting daubed by a blind artist.

"Smartness," he stipulates, "is the measure of what you see as *obvious,* what you can see as *obvious in retrospect,* what you can *invent,* and what you can *comprehend."* This in turn is a function of your "semantic primitives (what is simple in retrospect)." A Power's abilities might be so great that what it perceives as obvious, what it can "chunk" as a single conceptual atom or unit, will be the wholesale structure of today's human knowledge. Yudkowsky tries to catch the difference in kind that the Spike will create and represent:

> Think about a chimpanzee trying to understand integral calculus. Think about the people with damaged pathways from the visual cortex who cannot remember what it was like to see, who cannot imagine the color red or visualize two-dimensional structures. Think about a visual cortex with trillions of times as many neuron-equivalents. Think about twenty thousand distinct colors in the rainbow, none a shade of any other. Think about rotating fifty-dimensional objects. Think about attaching semantic primitives to the pixels, so that one could see a rainbow of ideas in the same way that we see a rainbow of colors.

One can appreciate Yudkowsky's sense of urgency, his passion for staring into the Singularity, while seeing the risks of such exultation: the banality always lurking as a temptation at its core, the ease with which prophetic wish might replace reason. Yet pause for a moment longer to consider the possibility that we might be able to stay alive long enough to share in this redemptive, or at least expansive, estate.

Whether by uploading or amplification, ordinary human beings—you and I, or our children and grandchildren—might move step by step into that condition. Into the Beyond, and then beyond the Beyond. Imagine your thoughts not merely accelerated but moving forward by immense jumps. Imagine how it might be if your feelings, your deepest emotions, were enhanced, your gifts of communication transcended as lavishly as an animal's grunts were transcended by speech and writing. It is an extraordinary prospect.

It is, in fact, the prospect of life after the Spike.

8: Earth Is but a Star

I THINK WE'RE MIDWAY IN THE CHAIN OF BEING FROM MICROBE TO
MEGAMIND, A TURNING POINT BUT NOT AN ENDPOINT. WE ARE A
TURNING POINT, AMONG OTHER REASONS, BECAUSE OF OUR TECHNOL-
OGY: WE ARE THE FIRST ORGANISMS TO LEAVE THE PLANET, TO DIS-
COVER FUNDAMENTAL LAWS, TO TINKER WITH OUR BRAINS AND
GENES. BUT THIS IS SURELY ONLY THE START OF THE AUTOEVOLUTION-
ARY PROCESS. I WOULD NOT EXPECT IT TO STABILIZE UNTIL WE AR-
RIVED AT, SAY, A GALAXY FULL OF JUPITER-BRAINS, ALL BENT ON
PROJECTS THAT WOULD MOSTLY BE INCOMPREHENSIBLE TO US . . .

WE ARE OF COURSE NOT BEYOND DEATH YET; WE ARE STILL RESI-
DENTS OF A DANGEROUS, PRIMITIVE PLANET. AND SHEER INCREASE OF
TECHNOLOGICAL POWER DOES NOT EXACTLY GUARANTEE UTOPIA. IT
DOES, HOWEVER, OPEN POSSIBILITIES, SOME OF WHICH MAY BE DESIR-
ABLE. FOCUSING ON THESE IDEAS NOW RATHER THAN LATER MAY DE-
TERMINE WHETHER THE SINGULARITY OCCURS WITHIN A CENTURY OR
WITHIN A DECADE. THE SINGULARITY MARKS THE HISTORICAL DIVI-
SION BETWEEN THE LAST TERRESTRIAL MORTALS AND THE FIRST COS-
MIC IMMORTALS; WHEN WILL IT BE? AND WHICH SIDE WILL YOU BE ON?

—MITCHELL PORTER,
"TRANSHUMANISM AND THE SINGULARITY," 1996

Ours is surely an unprecedented epoch.

Four hundred years ago, Galileo used his primitive telescope
to ponder the flawed faces of sun and moon, to spy out the
satellites of Jupiter. Now the corrected mirrors and upgraded
computer in the Hubble Space telescope are penetrating to the
very edges of the visible universe, and the orbiting Chandra X-
Ray telescope has resolved black holes inside and beyond our
galaxy. By the close of the first decade of this new millennium,
we might know at last exactly how the universe began and how

it is destined to end. (Of course, post-Spike humanity—or, more accurately, posthumanity—might intervene to reshape that deep future destiny.) Even ahead of the Spike and its promised down-pour of lavish knowledge, astronomers and particle physicists already have their crowbars well into the rusty locks of the universe.

As much-touted experts assure us, even before the completion of the Spike and its jolting plunge into transcendence, we confined humans already find ourselves on the doorstep of a valid Theory of Everything. Antagonists mock such a quest, pretending that this enthralling research really is expected to answer every conceivable question, magic equation in hand. This is a silly travesty. Scientists are an amusing lot, often with a Monty Python sense of fun. When they say elementary particles such as quarks have charm and strangeness, it's a pleasant whimsy. When they anticipate a theory of *everything*, of course they only mean everything relevant to their limited domain—the primary particles and forces of the cosmos. Ask for a burger with the works and you don't expect all the food in the store, plus the kitchen sink, to be jammed between the sesame seed buns.

So nobody need take fright when scientists declare that we really are, at last, with a stiff wind behind us, on the verge of the Ultimate Theory, once called superstrings, now known as M-theory. It has been invented (or perhaps discovered, ingeniously, painfully unfolded) by hundreds of scientists around the world, linked by the Internet. Its tutelary deity is Dr. Edward Witten, a sublimely gifted virtuoso in this mathematical art form. In the mid-1980s, he and his colleagues announced that the laws of quantum theory and of relativity—until then both fabulously successful but mutually at odds—could be fused. The haunting step in this unification was to posit that the world we know is a kind of shadow, cast by a far richer realm requiring not just four but ten dimensions within which things vibrate and interact.

Einstein had long since shown how gravity is just the bending or compression of width, breadth, length, and time, but you couldn't fit electricity and more mysterious nuclear forces into that sublime account. By adding extra dimensions, the equations clicked into place. Elementary particles, the stuff we're all built from, turned from dots into wiggly lines—pieces of one-dimensional string.

That was all very well, if hard to swallow. Alas, despite the extreme elegance of the formidable maths involved, at least five quite dissimilar stringy candidates emerged. Then in 1995, Witten announced a second string revolution. Adding one extra spatial dimension, he showed that perhaps all five models are views of the same profound, mysterious reality. Now it seemed that the basic objects of the world were not just vibrating strings—different vibrations appearing to us as specific particles and forces—but sheets or membranes (the M in M-theory). Shockwaves ran through the scientific community. Former skeptics, many of them canny Nobelists, started nodding, impressed. The quest surged ahead.

I follow this astounding and triumphant progress from the bleachers, but the view has been obscured by the extreme depth at which these specialists are obliged to work. Once physicists start talking about ripping and rejoining higher-dimensional objects in Calabi-Yau space, the phrases start to bathe you like the technobabble in *Star Trek* or *Babylon-5*. The electrifying difference, however, is that this time the hyperspace is not freely invented. We are on the verge of learning definitive answers to those ultimate questions that have driven curious, longing humans to distraction for tens of thousands of years. Where have we come from? Where are we going? Some of the answers that have been suggested by scientific research and theory imply that life after the Spike will just be the start of something far more majestic.

The cosmic Spike

According to the posthumanist theories of two mathematical physicists, Freeman Dyson and Frank Tipler, once life arrives in the universe it will find ways to persist forever. Stars will perish, black holes evaporate, but suitably adapted life will continue indomitably. Even after the last starlight has ebbed away into the blackness, Dyson argues (positing a cosmos doomed to endless

open expansion), life of a kind will prevail. It will be an inconceivably slow, electronic kind, for everything except electrons and
photons will be gone.

For Tipler, who hopes that the universe will close into a hot
Big Crunch inferno where everything is crushed together in a
reverse of the Big Bang, living beings will continue drawing energy from distortions in spacetime itself. For both scientists, life
must eventually fling off its physical hardware. And, if Tipler's
right, it will continue, in effect, forever—even if the spacetime
universe itself expires.

Current astronomical evidence suggests strongly that the universe is expanding *faster* as time goes on, due to what is termed
the Lambda field, which would invalidate Tipler's Omega Point
theory. This vast Lambda store of expansive energy, also known
as the cosmological constant, is now thought to comprise some
seven-tenths of mass energy in the cosmos, and to be blowing
everything apart. Tipler, though, is sticking to his guns for the
moment. In correspondence with Michael Shermer, of *The Skeptic*, he argued in 1999 that observed expansion rates might be
local rather than universal. After all, he declared rather obscurely,
he finds himself obliged to accept the Omega Point postulate,
since "the known physical laws leave me no choice . . . if the universe were to expand forever, either because it is open or it is
accelerated by a positive Lamdba, then Hawking has shown that
unitarity would be violated. But unitarity is one of the central
postulates of Quantum Mechanics, confirmed again and again by
every experiment to date."[1] This is a subtle objection from an
immensely sophisticated fusion of quantum theory and relativity,
disciplines in which Tipler is a world authority. Still, it is fair to
say that few other authorities concur with his analysis. They have
less at stake, of course. For Tipler, that's nothing less than his
eternal fate.

Mind, as we've seen, can be understood as the software of
advanced life, the programs of the brain. The crucial feature of
programs is that they are not necessarily restricted to one support
system. One data-processing structure can be emulated by another. A program may be run, with a few adjustments and some

changes to the code, on any suitable platform or substrate. If the universe plunges endlessly into the cold night, minds will slow but perhaps never quite flicker out. Conversely, if it crashes back toward the Big Crunch singularity literally at the end of time and space, as Tipler suggests, the communicating post-Spike computer minds of that epoch will accelerate to ever sweeter ease. If he is correct, they will be gods, smearing experienced time into an eternal Edwardian afternoon of divine reflectiveness.

Professor Tipler started his journey into deep futurity with an attempt to establish, purely by logic, the rather old-fashioned notion that human intelligence is alone in the universe. This unpopular theory blends seamlessly with a sort of revised and mind-boggling version of Human Manifest Destiny—save that Destiny beckons not to *us* but to our post-Spike silicon descendants, the Turing-cum–von Neumann automata we might launch into the heavens as our ageless deputies. In this view, we are just the precursors of the true inheritors of the cosmos. Those heirs of space and time will be the sublime constructs we now call, in our provincial pride, "artificial" intelligences.

Home alone

In brutal précis, here's the argument for believing humans might now be alone in the universe, or at least the galaxy. Start by supposing, for argument's sake, that life is common in the galaxy. A good proportion of the many fecund worlds predicted by optimists like Carl Sagan must have civilizations a few million years more advanced than our own. Any culture with an ounce of grit and curiosity will build at least one artificially intelligent, self-replicating starprobe. After a few million years, by natural increase, the galaxy will be rife with its offspring.

Since it takes only one culture with gumption in a universe of wimps to get the ball rolling, we should see such a probe in the sky. We don't, of course. Therefore something's wrong with the argument. The weakest link is the assumption that life is prevalent. So the absence of evidence for widespread interstellar civilization strongly implies that we're alone in the galaxy (or at the very least ahead of the pack).

If that assessment is faultless, current radio telescope searches for extraterrestrial intelligence (SETI) may be a waste of effort and resources. Smart life elsewhere in the galaxy would be unable to hide every last trace of its presence. We ourselves, for example, have already been broadcasting powerful radio and TV messages, inadvertently, into an expanding sphere two hundred or so light-years across.

Cosmic engineers

Even more strikingly, after the Spike we can expect that posthumanity may indulge in truly gargantuan works of engineering. Beings of such power might mine stars for antimatter, change their courses, ignite them to supernovae, and then snuff them out so that their new neutronium cores might be used for massive data storage.

So, unless we are the first kids on the technology block (an absurdly unlikely chance if life is widespread), it is a good bet that we are alone—at least in this galaxy of hundreds of billions of stars. That implies an awesome responsibility. Perhaps we are the fluttering candle of all future life in the universe. The sooner we share the spark the better.

Evolutionist Stephen Jay Gould has caustically criticized the argument, remarking that he has trouble enough estimating the motives of close friends without trying to second-guess the impulses of alien creatures. I suspect Gould's being naughty here, fudging levels of explanation. Just because you can't read your pals' minds doesn't invalidate the credentials of the sciences of psychology and game theory, used appropriately. The no-alien-life argument remains compelling in the light of current applications of game theory to evolution, which suggests that life is powerfully driven to reproduce and fill every available ecological niche.

God out of the machine

Certainly Tipler's vision is remarkable, for he suggests not only that life will persist and spread but that it will transcend itself, in the very deep future, into what he calls an Omega Point consciousness—a kind of immanent god, effectively all-knowing and all-powerful, yet evolved from the material universe. Here is its apotheosis:

> [I]f life evolves in all the many universes in a quantum cosmology, and if life continues to exist in all these universes, then *all* of these universes, which include *all* possible histories among them, will approach the Omega Point. At the instant the Omega Point is reached, life will have gained control of *all* matter and forces not only in a single universe, but in all universes whose existence is logically possible; life will have spread into *all* spatial regions in all universes which could logically exist, and will have stored an infinite amount of information, including *all* bits of knowledge which it is logically possible to know.[2]

Whether such eschatological fantasies have any place in the workshop of science is tricky to assess. An almost inevitable extension, one entirely salient to our exploration of the coming Spike and what lies beyond it, has also been developed by Tipler. He has gone outrageously further, suggesting that his Omega Point AI deity will fetch us back to life at the end of time, reconstructing each one of us, in a virtual universe inside Its own Mind, by the sheer power of its calculating prowess.

The power fantasies of science

This style of argument has not gone without intelligent and deeply skeptical critique. Moral philosopher Mary Midgley, for example, finds it no better than "self-indulgent, uncontrollable power-fantasies."[3] Unfortunately, Midgley thinks scientists like Tipler, Davies, and Barrow argue parochially from the existence and destiny of *Earth* life alone. Our world, however, has had a highly arbitrary evolutionary history, punctuated by accidents (such as a random rock from space wiping out the dinosaurs). No, Tipler and others are pointing to the gross improbability of many basic "settings" of universal physics, fundamental values of physical interactions that underwrite the very existence of stars and the elements we're made of. These happy features are needed by life of *any* kind. But while Midgley misses the crucial universality of the argument, her caustic reflections are quite compelling. I once asked Paul Davies, who has written favorably of the anthropic principle, if this gave him pause. Not a bit. Midgley, he replied sharply, "is beyond the pale."

How, though, is one to know what's within the scientific pale, in an era when some experts claim to find God in their equations, while others (such as Sir Roger Penrose) maintain that consciousness lies *beyond* the reach of currently known physics? Tipler's physical eschatology—traditionally, the theological study of "last things"—fails if today's physics contain some major hitch. Yet we know that science keeps bursting its own boundaries and gobbling up earlier certitudes. Tipler boasts, "I shall make no appeal, anywhere, to revelation. I shall appeal instead to the solid results of modern physical science [and] to the reader's reason."[4] While this is true, his post-Christian intuitions have obviously been shaped by existing religious templates, allowing his powerful physics to generate "a testable physical theory for an omnipresent, omniscient, omnipotent God who will one day in the far future resurrect every single one of us to live forever in an abode which is in all essentials the Judeo-Christian Heaven."

Midgley sarcastically dubs this "escalatorology": a fatuous belief in "an endless evolutionary escalator exalting the human race." No doubt she would excoriate the Spike possibilities we've examined in this book. Tipler's Omega Point divinity, too, is the child and culmination of human ancestors, thinking infinitely many thoughts infinitely fast in the very moment that the universe plunges into the Big Crunch at the end of time. It is a requirement of his theory that the universe end in fire, not in endlessly expanding cold.

This God is a kind of cosmic computer, written into the Higgs field that is thought to permeate the universe and lend particles their mass.[5] Given its effectively infinite capacity and benevolence—as proved, Tipler assures us, by game theory and microeconomic analysis!—this machine deity reconstructs and renovates us all, building us a virtual-reality paradise to enjoy for all of virtual eternity. So we'll come back as computer emulations of ourselves, but nicer. Even Hitler and Pol Pot, but they'll have to go through a lot of Purgatory run-time first.[6]

Transcendence without waiting for the Omega Point

Actually, we (or our transhuman children or great-grandchildren) could attain such a condition long before the end of the universe. Preserved by the wiles of science into the Spike, would we any longer be human?

Perhaps not, or not for long. Maybe we will live almost infinitely accelerated lives within a virtual computer in a grain of sand at the edge of the world's last drained sea. Or maybe we will be quantum states of a cosmically dispersed, radio-linked hypermind. Or maybe we will be, well, quite literally . . . *gods* . . . inflating fresh universes out of the quantum foam and placing our impress upon everything that forms there. Is that possible? Barely, according to an October 1999 *Astrophysics* paper by J. Garriga, V. F. Mukhanov, K. D. Olum, and A. Vilenkin, "Eternal inflation, black holes, and the future of civilizations."[7] If we do inhabit an inflationary cosmos, especially if its expansion is driven ever faster by a cosmological Lambda constant, new inflationary bubbles will "begin to nucleate at a constant rate. Thermalized regions inside these inflating bubbles will give rise to new galaxies and civilizations. It is possible in principle to send a message to one of them. It might even be possible to send a device whose purpose is to recreate an approximation of the original civilization in the new region." Such nascent universes, alas, would probably be cut off at the pass by black holes. Still, if sufficiently large negative energy densities prove not to be forbidden by quantum gravity physics, there is "hope for the future of civilizations." Doomed universes might fade into night, but their children will be breathed into the void and repopulated by their parent's informational seed.

Or maybe we'll all stay at home and watch the ultimate artificial reality television channel, forever.

Might this explain the Great Silence in the skies, the mysterious absence of any radio or optical traffic between worlds out there in the galaxy and beyond? If the technology and culture of every civilization Spikes within a century or two of discovering radio and space flight, they might all be tucked away into the folds of local spacetime like the hidden, rolled-up dimensions of string theory. They might even be living *there*, colonizing those intricate, infolded spaces.

The shape of Spikes gone by

On the other hand, is the universe itself already the reshaped handiwork of Spiked civilizations that preceded us? The cosmos has been around for between thirteen and fifteen thousand million years, and our planet only solidified some four and a half billion years ago. Subtle arguments suggest that the early cosmos must have been barren of carbon-based life for a very long time because, as we've noted, most of the elements were not created in the Big Bang but cooked inside exploding supernovae and seeded slowly through the gas clouds of deep galactic space. Still, statistically it would be surprising if we happened to live on one of the very first worlds capable of sustaining life and intelligence.

If the prospect of the post-Spike condition sketched in this book is correct—awesome, truly godlike powers wielded by Powers—then in another billion years our descendants will surely have plaited the stars into braids of their own design, if they wish to.[8] Tipler and Barrow, as we've seen, argue that an AI deity might, even must, emerge in the final nanoseconds of the Big Crunch after its predecessors have redesigned the dying cosmos. By reverse engineering, can we gaze outward now and see that the stars *already* bear the marks of cosmic engineering?

The Very Fast Evolution Machine

Here's an even more startling conjecture. Tipler's Omega Point deity, plunging into the forever of infinite compression, has an effectively infinite number of discrete clock ticks within which a "god" may Do All Things. But similar conditions existed during the initial 10^{-43} of a second when time and space were smeared together, or so Stephen Hawking assures us. Might life of some quite different ilk have crystallized in the strange, terrible epochs before our kind of matter settled out in the inflation rush of the expanding cosmos?[9]

In the earliest zillionths of the Big Bang eruption, time was effectively multiplied to infinite speed, but it slowed fast as spacetime expanded and cooled. Inconceivably vast numbers of force-particle exchanges occurred almost instantly in a densely compacted and connected spacetime where the four known forces of physics only "slowly" decoupled from a unitary force now lost forever, unless there's a Big Crunch at the end of the universe. Might not there have been virtual time enough, effectively, for a superintelligence to evolve from scratch? Even a whole bunch of them, but perhaps they would inevitably remain merged in a swarm-mind until the cosmos was big enough for light-transmission delays to disrupt module communication . . .

It is a suggestion that eerily resembles the teachings of the ancient Gnostics, in a way. The Gnostics held that our world is not the creation of an original supreme deity, but is the rather botched handiwork of a lesser god, a *demiurge*. Imagine not one but many angelic demiurges, the first-evolved minds in our cooling universe, tumbling from the furnace of the Big Bang, cast out into the freezing dark. Perhaps placing their impress upon the new regimes of matter and light. Yes, *now* there is a god . . .

But, if so, that was *then*. What of *today*? Would such "angels" still have any impact on the universe?[10] Would their works persist in the fabric of spacetime? The galaxies extend into space in colossal strings made of billions of stars wrapped about dark bubbled gaps, an arrangement that deeply puzzles cosmologists. Might this strange architecture be the remnant of some ancient design of the earliest life born of the Big Bang? More to the point, is the evolution of such "angels" remotely possible in the light of current physics?

Could any kind of high-level structure emerge under such appallingly volatile conditions, however many virtual steps or epochs it contains? It's one thing for life to persist into a Big Crunch, as Tipler proposes, using "shear energy" (the gravitational ebbs and flows of shockingly twisted spacetime). Presumably it's quite another for complex "life" to bootstrap into existence from nothing under the same conditions. *Or is it?*

Mitchell Porter agrees that the main barrier to Big Bang superintelligences is the absence of structure in the fantastically hot primordial plasma. "But conceivably," he notes, "there may have been epochs of structure in the course of the many phase transitions which are part of modern cosmological models of the early universe, and perhaps things were evolving rapidly enough for replicators to evolve." That catches it exactly. The contrast has been pointed up by Charles Stross, a British writer and software specialist: in Tipler's scenario the pre-Omega entity deliberately sets up oscillations in the collapsing universe, extracting usable energy. But did the Big Bang have equivalent energy gradients, available to drive such computational processes?

The cosmos shortly after the Bang was a homogeneous soup of radiation looking the same in all directions, Stross notes. On the other hand, the Cosmic Background Explorer satellite (COBE) detected ripples in the background radiation that suffused the universe. These are the enduring traces of lumpiness left in the pervasive radiation residue from the Big Bang. Later data from the COBE satellite suggested that they are, indeed, fractal in nature, ripples within wrinkles—perhaps enough to provide the gradients necessary to jump-start a primordial replicator.

The earliest ages of the universe

The opening fractions of a second in this universe contained ample variety. "GUT Age, Quark Age, Hadron Age, Nucleosynthetic Age, Plasma Age, Fireball . . ." Jonathan Burns, a La Trobe University computer scientist, suggests with a certain whimsy that, given these phases, "the blindest watchmaker would have had opportunity enough." He adds:

> What are the odds for an intelligent ontology?
> On Darwinian grounds one seems to need:
> (1) A substrate stable enough for some Selfish Form to persist and multiply in competitive variation.
> (2) A phenomenon which can be coded, and decoded, into a genotype which replicates the code.
> (3) Time for enough iterations that the code space can be explored by the population, long enough to find the breakthrough points to higher organization.
> (4) Time enough for the higher organization to explore its environment, and exploit the opportunities for technological enhancement.
> (5) A radically uncertain measure of good luck.

And Burns took up the idea of ancient demiurges with a poetic burst of his own: "The Benefactors . . . skating the contours of zero tidal force . . . their wingtips deep in blazing quicksand . . ."

Could such a selfish code-string persist though the fires of the Big Bang, and in the cooling cosmos left as its ashes? For a selfish signal to survive in a sea of noise, it has to perform its own noise reduction. Emergent exotica might stabilize briefly—vortices, frequency bands, phase boundaries—to form a first substrate. Efficient signal self-replication would use digital encoding, the simplest possible but rich enough to do the job. After all, we know that populations of data structures inside computers can already evolve, exploring combinational spaces efficiently, turning combinational complexity to advantage.

Is this kind of digital evolution plausible for the primordial universe? "The bulk properties of Grand Unified Theory plasmas are speculative, to say the least," Burns notes. "Electromagnetic plasmas, yes, there are stable structures, Alfven waves, in the right conditions. And in cold bulk matter, we get quantized magnetic flux tubes, and liquid-helium quantized vortices."

Physics has only vague ideas of how quark-gluon plasmas might behave. "One place to look for a clawhold might be at the point where the quark-gluon plasma is breaking into clusters. In the 'big bag' of the plasma, one gets incursions of vacuum, which acts as a superconductor for color-charge. For a sufficient epoch, just maybe the plasma is riddled with quantized chromodynamic flux-tubes in bunches. Asymmetry. Structure. Bistability. Gates and switches. Chemistry. New tubes being generated all the time, those which don't match *our* patterns discarded, the rest assembled into new entities."

Similarly with a conjectural breakup of the GUT plasma, or the compaction of the hidden dimensions. Emergent novelties, as Nobel laureate Ilya Prigogine argues, are often found at phase boundaries where energy is being exported into the environment. But still we wouldn't expect to find an *infinite* number of successive phase changes from the Bang to very shortly afterward, the literally uncountable sort required for a Tiplerian scenario. At the smallest scale at the start of time, quantum theory tells us, we would find everything-at-once, space and time smeared together and confused. If you can't count the ticks or intervals between each event and the next, or determine one place from another, it is impossible to create the structure needed for an intelligence.

Still, an infinite number of steps might not be required. After all, life has evolved and flourished on Earth in less than four billion years—quite a lot of separate clock ticks, but a good deal fewer than infinity. The ancient minds might have evolved and left their mark.

Traces of primordial engineering?

What legacy might such demiurges leave for us to find? It could range from the very large, such as cosmological gravitational waves, or the very small, such as strange matter in pulsars. "If the angels broke through to the mid-range, they could build just about any material structure," Burns comments. But is there anything in our stellar environment that can't be accounted for by available science? Well, there remain those mysterious cosmological features, the vast empty voids, and the so-called Great Attractor that appears to be dragging all the local galaxies toward a particular place in the heavens. And dark matter, up to ninety percent of the mass of the cosmos, remains an unsolved question.

"If I were an angel," Burns remarks wryly, "I'd be inclined to look out for my own skin. Maybe I could replicate myself on the cooler, rarer strata of the heat death. But in my epoch, the alternative of forming exotic black holes and maybe impressing myself on a new universe, if that's possible, would seem a lot more practical than it does to us atom-age relics."

Still more delightfully bizarre is a playful conjecture based on Tipler's cosmological deity, advanced by Anders Sandberg:

> life evolves toward the Omega Point, but in the vicinity of the final moment "angels," life based on back-propagating causality (which Tipler's theory seems to imply) are created and move backwards through time. They are unobservable in the present, since they are acausal from our perspective . . . and probably *very* thinly spread (possibly "extinct"). Eventually conditions become better and better for them, they spread across the universe and use the shear energy to create the Alpha Point—which is isomorphic to the Omega Point and creates "angels" moving forward in time. Note that if the backwards-moving beings use shear energy from the "collapse" of the universe they see, this may explain the homogeneity and isotropy of the universe despite the chaos of the Big Bang—from our perspective they smoothed the early universe!

In terms of scientific cosmology this entire arabesqued line of thought is strictly unnecessary, since science does not lack in more modest explanations for its outstanding conundrums. Still, improbable as it is, it does bear a piquant resemblance to the issues that might arise when Powers in a post-Spike history start to reformat their virtual and real environments.

So far, this territory has been the stamping ground of amateur gamesters like Burns, Stross, Porter, Leitl, and Sandberg, and scientifically trained speculators such as Stephen Baxter, Gregory Benford, and Greg Egan. Perhaps the leading academic explorer of runaway change, of the lead-up to the Spike and of post-Singularity possibilities, is Hans Moravec, the robotics specialist at Carnegie Mellon.[11]

The age of minds and robots

Moravec's guided tour from the near future to something approaching Tipler's Omega Point universal Mind, in his 1999 book *Robot*, is at once exhilarating and daunting. As Yudkowsky points out, it is really impossible for us to anticipate the nature and behavior of deep-time Powers. Moravec's own time line, however, boldly extends from the short run (early 2000s) through the medium term (around 2050) to the long run (after 2100), and thence into the voids of cosmological future history.

The near future – robots everywhere

Automation and robots, already with us, will replace human workers in ever greater numbers, and at ever loftier realms of subtle expertise. We swiftly reach a point of absurdity. Prices for consumer and other goods will fall swiftly, as the costs of making them diminish—and as the income of consumers vanishes when available paid work dries up. We can foresee a ludicrous era in which AI-run factories would stockpile vast mounds of goods that we human consumers can't afford to buy.[12]

Downsizing (as sacking workers is euphemistically called these days) saves corporations so much money—eventually, all of their labor expenses—that nobody except the stockholders will have any income, except what is redistributed by central agencies, especially tax-levying governments. One way around this impasse is to encourage stakeholders to buy stock early, as happens now in high-tech companies such as Microsoft. A slice of the future action is thus part of the remuneration package. It's a policy that has turned many Silicon Valley employees into millionaires. Besides, in some respects we already approach this condition, since pension funds already hold much of the capital in advanced economies.

The medium future—a tribal utopia

These problems will be solved—radically, because that's the only way to do it. But not violently, or through ideological revolution. Citizens of the automated (or early nano minting) First World might resemble those of today's Middle East petroleum kingdoms and emirates. Oil's black gold has paid for expansive new cities, excellent education, free health care, and imported labor. As for the Swiss, life for the majority is comfortable, relaxed, prosperous, and untraumatic. That might change with nanotechnology, when the poor fit between luxury and Islamic doctrine comes into bloody focus.

Near-Spike technologies will ensure that everyone could have access to this cornucopia. While different cultures and moieties within those cultures will respond idiosyncratically to the great change, we might expect a general drift back toward the ancient behavioral codes implanted by evolutionary pressures. Humans evolved in a million years and more of moderate ease, each tribe's small numbers drawing modestly upon the self-replenishing bounty of the earth. We will slip easily back into that blissful state, spared its occasional lethal bouts of uncontrollable fire, flood, drought, plague, and infestation.

Since our ancestors had to survive such disasters, we evolved with reserves to meet extreme demands. When need drives us, we readily exert ourselves. Developing herding, farming, and machine technology has nudged us, step by insidious step, into acting as if that emergency prowess were the way we ought to live all the time. The results are sometimes lethal stress, anger, road rage . . . Tribe finds itself pitted against tribe. By the middle of the twenty-first century, however, with the work fetish vanished willy-nilly, it might be the best features of tribalism that rise to save us from ennui and the stress of nothing to do with our jobless time.

Those in the West may "return to a comfortable tribalism," Moravec predicts, "after a five millennium detour into organized civilization. Countries with traditional tribal structures may simply stay that way, building on their ancestral customs, leapfrogging urbanization altogether, while developed countries foster tribes with customs and beliefs more bizarre than anything today" (*Robot*, p. 137). In my view, the cycles mapped by the last thousand years of Western history suggest that we shall pass in any case through a period conducive to recovered tribalisms and local loyalties in the first half of the century, so this prediction gains support from several independent lines of thought.[13]

Controlling the AI workforce

How are those vast automatic or artificially intelligent forces of production to be contained and directed? Company law might be our best model, Moravec suggests. Certainly robots, even robots with some kinds of mind or consciousness, need not become franchised citizens. The issue of AI slavery is not as sharp as upload piracy or servitude, because the design of robot minds is to some degree under our control. Even if that grows less and less true—since those hypercomplex mentalities will be forged by genetic algorithms competing in computational space under the pressure of Darwinian contest—still we can ensure that they *enjoy* doing our bidding. That would be vile if applied so bluntly to human education (although a moment's cynical reflection might lead one to wonder if, after all, it isn't already education's design), but is less obviously so for minds shaped from the outset to a task. For all that, rogue AIs might escape any corporate laws installed (like Isaac Asimov's fictional restrictive Laws of Robotics) in the grammar of their behavioral repertoires, so we'll need "police" clauses and something like antitrust laws to prevent a wild mentality growing ever larger and more powerful.

At this point, we start to lose any vestige of our commonsense grip. As AIs rewrite themselves and accelerate their own powers (commodious memories, wider cognitive windows to juggle more factors simultaneously, swifter processing pace), they will grow wilier as well. Just as an infant understands its own urges and frustrations less well than its parents do, AIs will *see right through us*. That is the humiliating relation in which we'll stand to them, even at the level of comprehending our *own* motives, let alone theirs. "Corporations struggling to appeal to consumers will develop and act on increasingly detailed and accurate models of human psychology. The superintelligences, just doing their job, will peer into the workings of human minds and manipulate them with subtle cues and nudges, like adults redirecting toddlers" (p. 142).

This is an appalling prospect, unless you expect to splice your awareness into one of these supermachines, or upload into their company. The unaugmented human citizens of Earth might find themselves obliged to create a new protocol:

To exceed the limits, one must renounce legal standing as a human being, including the right to corporate police protection, to subsidized income, to influence laws—and to reside on Earth. In return one gets a severance payment sufficient to establish a comfortable space homestead and absolute freedom to make one's own way in the cosmos, without further help or hindrance from home . . . Freely compounding superintelligence, too dangerous for Earth, can blossom for a very long time before it makes the barest mark on the galaxy. [p. 143.]

That option, as you can see, implies that the arena opens explosively at this point—off the planet, to the moons and local planets, and to colonies in orbit. That is the long-run framework for 2100 onward. It is an astonishing timetable, when you think about it, no more distant from our own era than today is from the earliest powered flight.

The long term—the Exes appear

Probably Moravec's version of the Spike, with its curves running almost straight up the graphpaper, occurs somewhere in the second half of the twenty-first century, or the start of the twenty-second. That is when we see something unprecedented: the birth of the *Exes*.

Each of the beings Moravec denotes as an Ex is indeed an X, an unknown (and, here and now, an unknowable) quantity. They are also Ex-es, formerly human or formerly AI-corporations, *ex*-traterrestrials once off Earth, so that they are now *ex*-iled to the dark between the stars.[14] But these *X File* descendants, Gothic space-age monsters in the Attic always overhead, will not truly be Ex-anything, or Neo-anything, for that matter. They will be *sui generis*, a new phylum of life, an order of organized matter and energy and information outside our categories. For now, however, let us agree to call them Exes, aware that the reality will snatch away any strong parallels Moravec or we care to build from our ordinary experience.

Exes in space

Exes are not a single species, but an entire ecology of endlessly diversifying species. In a way, each full-blown, self-reorganizing, self-replicating Ex will be a species in itself, as medieval angels were deemed to be. (The analogy is, once again, inevitable. After all, biblical Powers were an order of angels as well.)

As in life and A-life, parasites will burgeon, cropping the earlier Ex forms. Vast, intricate antibody systems, themselves ferociously intelligent, will roam the meme systems of each Ex, hunting down foreign predators and pests and eating them, perhaps absorbing their own unique structural contributions into the greater good of the Ex. That, after all, is how many useful mutations and genes or parts of genes are thought to spread in organic life-forms, either within or between species—and, if Sir Fred Hoyle's curious conjectures are valid, between worlds as well.

What kinds of handshaking or weapons-brandishing can we expect between such increasingly powerful entities? They will be fueled, after all, by the free flux of solar energy and the power of fusion reactors, drawing their material needs from barren rocks abundant in the moons, comets, and asteroids of the system, and hence effectively independent of each another. Moravec suggests, perhaps surprisingly, that the watchword will be *trust*.

For one thing, these beings will be so smart that not much will get past them. For another, they will have huge, and long, memories. A game-theory strategy called "Iterated Tit-for-Tat" is known to be a particularly effective and remarkably stable strategy in our world. You act in a kindly way until betrayed, and then punish treachery with swift reprisal, returning to kindliness the moment your foe offers it in turn—departing from this rule with spasms of random opportunism just often enough to throw cheats off guard. Repeated through many cycles, it leaves a window for sly defectors, before they are smashed in retribution, but in the very long run it seems to work better than any other scheme.

The Exes will probably play this game more adroitly than we do, even if they have not rewired their emotionality to the Mr. Spock coldness usually portrayed in pulp depictions of the superhuman. With a superb memory and a clear sense of the logic of interactions, no Ex is likely to bring down the concerted hatred and reprisal of the rest upon itself. After all, Ex communities remain bonded by a skein of radio and optical links. The AIs will exchange scientific and artistic creations, hire each other's skills, even purchase entire personality structures. They will be governed, even in deep space, by the laws of the marketplace.

At the same time, the Exes will be entities in constant flux, morphing their bodies and their minds. Perhaps they will follow the twofold principle found in organic creatures: a stable and protected coded *recipe* (written in DNA in life-forms, in computer code among the Exes) together with a flexible and fluid phenotype or physical *expression* of the recipe. A neat analogy Moravec offers from a nonliving memetic structure, the political nation, is the double structure of *constitution* (fixed against easy change) and *laws* (the practical interpretation of those principles, stable in the short term but readily recast to fit new circumstances). Or *beliefs* and *core ideology* versus *knowledge*; scientific cultures expect the latter to be endlessly in motion, self-subverting, expanding into unexpected domains.

The large and the very, very small

Exes will not necessarily be immense structures, Deathstars roaming the heavens. In a time of nanotechnology, their working parts will be as small as technically feasible. A human mind could be emulated power supplies, coolants, and all—in a cube a millimeter square. At the top end, mighty Minds could be as large in extent as a hundred-kilometer asteroid, but not much bigger, because speed-of-light signal delays would drastically slow access between its modules, forcing them to operate independently (although in coalition or parliament, like the diverse subsystems of the human brain and body).

Some transhumanists dream of uploading into AI brains the size of gas giant planets, "Jupiter-brains." That hope might never be realized, defeated by the speed of light—although, as noted earlier, a Neuron star might pack a cosmically vast amount of processing power into a spinning sphere of incredibly dense neutron matter the size of the earth. That would be, Moravec suggests, "a mind whose components are a million times as closely spaced and a million times as fast as those in regular matter. Like sages on remote mountaintops, isolated, immobile Exes trapped in neutron stars may become the most powerful minds in the galaxy, at least until other Exes accumulate stellar masses of heavier elements to build neutron computers in their own neighborhoods" (p. 162). Lesser Exes might resemble a core surrounded by a nano Utility Fog, each tiny component speaking to the rest by laser light: "surrounded by an illuminated cloud that does its bidding as if by magic" (p. 150).

Beyond the Exes and into the cosmos

Is this the true Spike? A further, barely glimpsable era will follow the epoch of the Exes. Now the cosmos gradually fills with a vast, replicating globe of Exes fleeing outward from the sun, consuming or subverting or assimilating everything useful that stands in its path. Exes will accelerate at nearly the speed of light, learning more and compressing that knowledge into deeper and subtler formulae.

Out of direct contact with their source worlds, but skeined across the cosmos in a sphere of information-dense communications, the Powers will learn how to reshape space and time to their own advantage. They "may learn to tailor spacetime at the finest scales into improbable meaningful states," Moravec speculates, "that are to common particles as knitted sweaters are to tangled yarn" (p. 164). The transhumanist Mitch Porter has dubbed this epoch the Second Singularity, the superforce phase of spacetime engineering capabilities.

And at that point, if not before, the universe will become a substrate for *datavores* (the memetic entities discussed at the end of chapter 6) that consume and fashion pure informational structures. Data eaters would be unconcerned with the crude forces and fields and particles and standing waves that they are encoded upon. Personal identity, always moot for post-Spike beings, will perhaps vanish altogether, as the physical locus of memories fades forever from importance. The dream of the upload theorists will be fulfilled: you will be where you are thinking, in as many embodiments as it takes, or as you ("you") see fit. Which suggests strongly to me that plans to meet up on the far side of the galaxy for the Bean Dip Party are quaintly misplaced—first, because you'll be there already, and second, because there'll be no "you" there.

Just as identity vanishes, so too will the usual metrics for space and time. Right from the outset of your life as an Ex, the enhanced and multiplying speed of your inner life will make everything else around you seem farther away, and it will appear that the universe is running at a slower rate. Add to this the quite objective dilatation of time when a spacecraft travels near the speed of light, and we see that life and mind will become "smeared out" across the history of the universe.

Worlds within worlds

Aside from these factors, there's the appalling physical shrinkage we've emphasized previously. Every advance in understanding leads to new technology, and every breakthrough in technique compresses the size of a machine's components. Finally, we'll have not just Shakespeare and the Bible written on the head of a pin, but a faithful miniaturized copy of the universe itself:

> Today we take pride in storing information as densely as one bit per atom, but it is possible to do much better by converting an atom's mass into many low-energy photons, each storing a separate bit. As the photons' energies are reduced, more of them can be created, but their wavelength, and thus the space they occupy and the time to access them will rise, while the temperature that can obscure them drops. A very general quantum mechanical calculation . . . leaves room for a million bits in a hydrogen atom, 10^{16} in a virus, 10^{45} in a human being, 10^{75} for the earth, 10^{86} in the solar system, 10^{106} for the galaxy, and 10^{122} in the visible universe. (*Robot*, p. 166)

I have no real sense of the magnitudes Moravec is describing here, and I doubt he does himself—such numbers exceed the grasp of unaugmented brains. But he does manage to make it slightly more comprehensible, in a truly shocking piece of arithmetical legerdemain. If a human brain, body, and supporting environment contain roughly 10^{18} bits of information, then the world's whole population might be encodable by 10^{28} bits. This is a dreadfully large amount, by definition a whole world of information. But look at the numbing implications: coded as a properly efficient cyberspace, the bits represented by the atomic particles of a single human body (10^{45}, as we've just seen) exceed the entire world's human population by an astronomically vast factor.

And this logical implication, however extreme, suggests that my earlier wild speculation about demiurges from the era of the Big Bang might not have gone far enough. Might not the whole universe, including the earth, the warming sun, our own bodies and brains, already be inhabited by waves of *kenes*, memetic entities that flow through us as electricity passes along wires?

Perhaps it is an untestable and therefore, finally, an uninteresting idea. If such higher-order beings do nothing to influence our lives, why should we care if they are there or not? Certainly they do not map neatly onto any ancient or modern theological doctrines. They are *not* tradition's angels, nor Gnosticism's demiurges, and certainly not Yahweh and Satan. Nor, I think, could they have anything in common with the imaginary and rather vulgar "higher vibrational" spiritual energy beings of Theosophy and other mock-Eastern systems of thought. But such skepticism might just be my limited imagination . . .

What happened to the other Spikes?

Again and again, our faces are pushed into a kind of cosmic paradox, mentioned earlier when we discussed Frank Tipler's vision of humanity's role in the cosmos. It's been dubbed the "Fermi Paradox": "Where are they?" asked the great atomic physicist Enrico Fermi, looking at the silent skies. Alien Spiked civilizations might be sensibly close-mouthed, fearful of others of their kind from alien stock, or indeed from their stock mutated by different histories.

All flesh is grass, saith the prophet, and all grass is food. If you don't wish to be eaten by someone else's mouth, you're well advised to keep your own buttoned tight. Which doesn't mean that absence of evidence is evidence of absence. They might be there, everywhere. They might be here. We just don't recognize aliens even when we breathe them in and out, or let them rush like a sigh through the atoms of which we are composed.

Perhaps the nearest to such an explicit perspective is the "dirt" theory suggested ebulliently, perhaps tongue-in-cheek, by Stephen Witham. "Any sufficiently advanced communication," he proposes, with a nod to Arthur C. Clarke, "is indistinguishable from noise." If, to the naked eye and naïve ear, much of the cosmos seems like sheer random jitter and clang, that might be no more than you'd expect of a high-grade encryption program.

Lately there's been a lot of fuss about ciphers and secrecy on the Internet. Using a protocol dubbed PGP, or Pretty Good Privacy, you can run your e-mail or business documents or cash transaction through a preliminary filter and turn it into a two-part jumble of letters and numerals that can't be unscrambled without your private key. Others can, however, use your *publicly* available PGP key to test whether a message purportedly from you actually does have your seal of approval. The neatest form of such encryption, much prized by extropians and other net libertarians (such as the programmer and lingerie model Romana Machado, otherwise whimsically known as Cypherella) is Steganography, which hides your message in the background of picture files. The profile or spectrum of a well-encrypted message, efficiently compressed, is perfectly "white," indistinguishable from sheer hiss or a random scatter of pixels throughout the image you transmit.

Witham had a nice idea. What if the universe we see is background-coded with the Minds of our betters, entities that long ago Transcended or Spiked? This would be a theory of Cryptocosmography, and its Monty Pythonesque maxim might be: *every grain of dirt is sacred . . . and perhaps watching you.* Witham put it this way:

> We don't know we're not looking at "alien" civilizations. We don't know that the whole universe isn't colonized. Life evolves to become efficiently-encoded information, which looks like sunlight and dirt. I think these are the most natural developments to expect. The default scenario. I would expect a colonized universe to look exactly like a barren one. So what was the Fermi "paradox" again?[15]

We should not expect to see a cosmos blazing with crude antimatter battles between berserkers—dedicated life-killers whose five-billion-year mission roaming the void is to seek out strange civilizations and exterminate them. No, Witham's fear is "interpenetrating infections, computer viruses in the kernel level of physics," a kind of "applied theology." If that's feasible, it might be that we already inhabit a universe entirely colonized at *all the interesting levels* by post-Spike cultures. That would be the mother of all dirty goo catastrophes. Except that it's not, strictly speaking, a catastrophe. It's just how things are. "At most, our civilization, life as we know it, is the faintest ripple, the merest wisp of a breeze, on what's going on right in our laps. We are an insignificant perturbation not yet worthy of scratching, information-theory-wise."

As you might imagine, this rude suggestion elicited baffled or angry responses from critics. Science *already knows too much* for this to be true. There's no room in physics for hidden gods lurking in the dirt, or in the atoms, or in the folded-up dimensions. Anyway, computer design is well understood, and data routing and bit exchanges don't look one whit like noisy dirt. Get out of here![16]

Others noted that, well, really we still only know a *teeny part* of everything that's yet to be known. Besides, the point is not that computations run to resemble noise are *efficient*, and therefore detectable, but that this masquerade of noisiness might be the only way to stay free of a bug-squasher able to stomp your star. (Not necessarily a big problem for post-Spike technology, but this is a debate for advanced game theorists.) "I imagine aliens with billion-year patience would have extra slack here," Witham noted, probably with a grin. And if aliens can be expected to comply with game theory to this counterintuitive conclusion, maybe tomorrow's post-Singularity Exes and their human pets will do the same. Our immediate and recognizable merely human descendants, if there are any who elect to refuse the uploading option, might end up living in a paradisal world exactly like Pleistocene springtime, eating of all the trees in the garden except the Tree of Knowledge . . .

Rewriting the cosmic laws

Polish polymath Stanislaw Lem once made a similar suggestion.[17] Then why don't we find all those archaic galactic civilizations?

> . . . because they are *already everywhere* . . . A billion year-old civilization employs [no instrumental technologies]. Its tools are what *we* call the Laws of Nature. The present Universe *no longer* is the field of play of forces chemical, pristine, blindly giving birth to and destroying suns and their systems . . . In the Universe it is no longer possible to distinguish what is "natural" (original) from what is "artificial" (transformed).

The primordial cosmos might have possessed different laws in different regions (a notion common to current claims by cosmologists Fred Hoyle and Andrei Linde). If so, only in certain remote patches might life arise. Attempting to stabilize its environment, each early Spiked culture would jiggle the local laws of physics to its taste, until in their hungry expansion for living space they begin to encroach upon each other's territories.

Vast wars would follow: "The fronts of their clashes made gigantic eruptions and fires, for prodigious amounts of energy were

released by annihilation and transformations of various kinds . . .
collisions so powerful that their echo reverberates to this day"—
in the form of the 2.7 degree Kelvin background radiation, mistakenly assumed to be a residue of the Big Bang. It is a charming
cosmogony—an explanation for the birth and shape of the observed universe—and it fits all too neatly with the colossal inter-
galactic filaments and voids first detected years *after* Lem
published his jape . . .

This universe of Lem's, torn asunder in conflict over its very
architecture by titanic Exes and Powers, is saved from utter ruin
by the laws of game theory, which ensure that the former combatants must henceforth remain in strict isolation from each
other. The chosen laws of physics that prevail, as a result, are just
those restrictive rules we chafe under today: a limited speed of
light chosen to slow conflicts, an expanding spacetime (good
fences make good neighbors, don't you know). We live upon a
scratchy board abandoned by the Gamers. The Universe observed
and theorized by science is no more than "a field of multibillion-
year labors, stratified one on the other over the eons, tending to
goals of which the closest and most minute fragments are frag-
mentarily perceptible to us."

This delicious logic was not a bid by the distinctly atheistic
Stanislaw Lem to reinstate a religious perspective in his then-
communist Poland—something that the triumphant revival of
Catholicism has done in the meantime, no doubt to Lem's chagrin. Nor am I seriously suggesting that this is how our universe
really began. But the scenario does sketch out rather brilliantly
just the kind of universe we might expect this one to *become,*
following the human Spike. If so, has it happened elsewhere already?

A perspective that professional cosmologists fail to acknowledge (I can see their faces screwing up already) is that the ob-
servable universe, in whole or part, might indeed be at least
somewhat engineered, but not by any known religion's deity. You
can see why they'd have little sympathy for that conjecture. The
Copernican Principle, which has served science well for centuries,

tells us that the safest default assumption is ordinariness, medi-
ocrity. Things just *are* how they seem. There's no immense neon
advertisement in the heavens informing us of the presence (or
departure) of cosmic civilizations.

But hang on. Certainly, we now suspect, there's been plenty of
time for other life-bearing planets to form, hatch their brood,
nurture intelligence, seed it into the cosmos at nearly the speed
of light (or much slower, it makes little difference). That's a logical implication of the same Copernican Principle. We humans
will probably follow this course sometime between the end of the
twenty-first century and a million years hence. So why should we
be unique in this respect alone?

If that's correct, our own galaxy with its 400 billion suns and
at least 10 billion-year history has had many opportunities to
bring forth Spikes aplenty in the heavens. True, the earliest stars
would have been deficient in heavy elements, but there have been
stars like the Sun for many hundreds of millions if not billions
of years longer than our own 5 billion-year-old star. What would
galactic colonizers look like when they're at home? Let us look
carefully not for lurid displays (which are boastful, immature,
tacky, and probably dangerous) but for clever husbanding of resources by one or more sublimely competent technological cultures scattering their mind children across the sky.

A Russian astronomer, Nikolai Kardeshev, charts a taxonomy
of possible cultures by their command of available energy resources. We're Type 0, just up from rubbing two sticks together
to make fire. Type I harnesses all available energy of a planet,
drawing on tidal forces and heat at the world's center, in a moderate social order united by a global communications network. A
Spike would transform us very swiftly from Type 0 to Type I.

A Type II builds colossal physical shells of solar radiation ex-
tractors around its central star, marshaling as many photons as
it can into its energy budget. Such a rejigged or cloaked star
would seem to wink out, radiating only dim "black-body infrared" light. Type III civilizations are true galactic cultures, linking
local and far star systems at the speed of light, transforming crude

stellar nuclear furnace energy into ever deeper orders of information. If there is a Type IV, it would bridge the billion light-year gulfs between galaxies.

Do we see evidence of such alien Spiked cultures? Not at the largest scale, it seems, since the heavens are still filled in every direction with brilliant starlight. (But bear in mind that beyond a certain distance, say six billion light-years, there might have been no star systems ready to harbor life. And once started randomly, life might take further billions to attain intelligence, as it did here. So perhaps we should look for key differences between stars near us and those farther away. Nearby stars are cosmologically "older," made of rich, recycled star-loam. Many of the farthest, "young" stars from early in the universe's history, just now being captured by our best telescopes, perished long ago. These are the paradoxes of observation in a universe where light-speed is the limiting velocity.)

It does not take long for an aggressively colonizing society to fill a galaxy. With nano and AI (technologies as inevitable as basic arithmetic, one would suppose) it will have little trouble converting every usable sun into a home for superintelligence, dwelling in linked communities of star-circling shells. So the fact that the skies are not altogether dark at night seem to argue either that the whole Spike proposition is nonsense, or that any culture venturing in a Spikish way wipes itself out before having much impact beyond its own world. Or maybe that we aren't thinking about this right.

Star dolls

Robert J. Bradbury is an energetic programmer with skills in diverse but relevant disciplines: cosmology, aging, neuroscience, genomics. He has proposed an audacious solution he dubs "Matrioshka Brains." Recall those cute wooden dolls, a bit like painted bowling pins, which open to reveal nested inside them a second doll—which opens up to disclose a third, then a fourth, then a fifth . . . Dolls all the way down. Imagine a stellar system of concentric Dyson shells using the same design. But these shells would not just be platforms for life, as Dyson imagined; they would *be* alive. "Advances in computer science and programming methodologies are increasingly able to emulate aspects of human intelligence. Continued progress in these areas leads to a convergence which results in megascale superintelligent thought machines" that "consume the entire power output of stars."[18]

This is not the place to pursue Bradbury's inventive and nicely thought-out analysis of life as an effectively immortal star doll. The interesting question is this: if Matrioshka Brains are indeed an optimal solution for all intelligent life-forms (including organic life uploaded and enhanced within these vast habitats of thought), what would the cosmos look like once they are common? Even more exhilarating: if they *already* are common in the galaxy (and we have reason to expect life to be plentiful, even if it only arises rarely and has to send out its seed), can we see them if we search the observational data? Suddenly it's no longer a matter of "is there any room in our cosmology and physics for them?" but "do we *need* them to explain what we see?"

Bradbury's elaborate calculations argue that a very good—perhaps optimally best—solution for an energy-husbanding superintelligence is an immense computational network nested in layers about a suitable star. He is remarkably conservative in his projections, refusing (like Frank Tipler) to permit wild or exotic physics unknown to today's science. That might turn out to be an error, but it is a methodological choice a scientist is almost obliged to adopt; otherwise *anything* becomes possible and no proposal is interesting or testable.

Restricted to known physics, then, you have the job of building a nested computer on a scale to beggar the imagination. Using nanoassemblers, first decompose asteroids and eventually lifeless planets into streams of elements, launching them into appropriate orbits where they are recompiled into solar power collectors and computers.

Start with a shell of mirrors and solar cells not too close to the sun lest its structures evaporate, say between Mercury and Venus. Turning this energy into information and storage creates order in one place but waste heat in another. That waste needs disposal, so it is radiated outward from the hot inner shell. But one critter's waste is another's dinner (as ecology tells us), and the degraded energy can now be trapped and put to use by a much larger outer shell, and the residue of that one by another farther out, and so on. It's a setup "designed to take advantage of the downhill thermodynamic slope," Bradbury observes. "Each layer of a 'Matrioshka Brain' harvests the energy (optical or heat radiation) of the next inner layer, performs what work is possible with that energy, and radiates it at an even lower temperature."

This makes good engineering sense, even if the scale of the job is mind-boggling. Luckily, abundant elements (including carbon, silicon, iron, and oxygen) can be used for building nanocomputers in the largest outer shells. The first portions of this construction job can be done with appalling speed, given self-replicating mints and an airy unconcern for heritage values. Pull apart a five-kilometer asteroid and build light-sucking solar collectors. Beam their energy at Mercury, powering a swarm of metastatizing nanodeconstructors, and the planet will be reduced to usable rubble in . . . no more than a *month* (if Bradbury's figures are right). While it might take many years to pull Jupiter to pieces, that project awaits the long haul. Eventually you have a star coated in layers of thinking machinery and power transformers. What does it look like from a distance? Can you *see* it?

That's the intriguing question. Obviously the entire radiation budget of a blazing star has to be retransmitted at last by the outermost shell—but now it is spread across a radiating surface at least as wide as the orbit of Earth and perhaps stretching as far across as Neptune's orbit, and recovered from the bottom of the thermodynamic cascade. So the star will look dim, exceedingly dim. Maybe so very dim that even if nine-tenths of the cosmos is already transformed into M-Brains, all you'll easily detect will be the influence of their stars' gravity. They will be . . . *dark matter*.

Ah. Just what the puzzled cosmologist ordered, you might suppose. In the last decade or so, there's been no absence of wild and woolly theories and models and exotic explanations for a visible disparity between the gravitation holding galaxies together and the shortage of stars to generate that immense field of force. Physicists juggle black holes and the cosmological constant and brown dwarf stars and undetected new particles, just to make up the needed mass. Bradbury offers a fresh candidate: a surfeit of Matrioshka Brains.

One convenient place to look for them might be globular clusters, aggregations of up to millions of stars bunched into a ball only hundreds of light-years across. A very desirable location, location, location if you are a star-sized intelligence (or coalition) restricted to light-speed communication with your neighbors. Is there any evidence of clustered M-Brains? Certainly, says Bradbury. Observations show large numbers of low-luminosity red stars, which is usually taken to imply old small mass stars. It could as easily be the sign of hot stars shelled by radiators. What's more, low metal abundance in galactic cluster light spectra implies "old stars formed before significant numbers of supernovae had seeded the stellar dust with metals. However, the low luminosity of stars in a [cluster] could be due to light harvesting and redirection for power and the low metal abundance due to metal mining for construction projects."

The objections are patent, and critics are quick to voice them. Why only some stars transformed this way, and apparently at random? (We don't see large patches of dully radiating sky, the mark of a vigorously colonizing culture. No, the "dark matter" is spread evenly throughout the brilliant glowing tracery of visible stars.) For that matter, Fermi's Paradox arises again. Why have *we* been spared? Our sun and yummy solar system might make a very fine foundation for someone's settlement and transformation. Bradbury offers answers to these challenges, and I think it's fair to say that while they are not finally persuasive he has opened up a peculiarly enthralling and shivery way of looking at the heavens. The prospect of his being wrong is equally pleasing, since his suggestions show us a possible path for a Spiked humankind and our posthuman offspring.

On the other hand, it's bracing, I suspect, to acknowledge in due humility that, for all we know, actually there *are* other Powers in the cosmos, right now, who have passed through the veil of the Spike. Perhaps their physics is to ours as ours is to Aristotle's, or an ant's; perhaps to them, a Matrioshka Brain is a relic as ancient and irrelevant as a trilobite shell. And perhaps they do move upon us, vast and heedless, as fire moves across the tops of a field of cropped and stubbled wheat . . . [19]

Meanwhile, here and now . . .

Let me drag you back from the sublime to the urgencies of the present pre-Spike crisis. According to a study by the International Labor Office, as many as a billion people in the world, this minute, are unemployed or underemployed. This is an appalling waste of human potential, as well as an unprecedented quantity of abiding human misery. On the other hand, except in a few running sores of dispossession and active cruelty, it may be a more bearable kind of misery than the suffering and uncertainty felt by the bulk of humanity throughout history.

The standard cure prescribed by dominant economic theory is *growth:* dig up more, plant and harvest more, process more, make more, sell more to more places . . . This solution, admittedly, did power the first industrial revolution, and certainly succeeds wildly even now in bootstrapping certain impoverished nations into a thrusting mood of international competitiveness (while leaving others in monocropped and exhausted ruin).

But beyond any debate over policy, as Theobald and a thousand concerned voices remind us, disastrous ecological crises are entailed by ceaseless industrial growth *of the kind available to us today.* Russian men, science writer Julian Cribb reported, typically die before retirement age. Worse, a "spreading epidemic of babies born with a sickly yellow hue, of rising mental retardation, deformities and acute chemical sensitivity, spotlight the extent of a crisis underpinned by collapsing water safety, choking air pollution, contaminated food and the ruin of the forests on a scale never before seen . . . Russia under communism was a notorious environmental black hole . . . but what has ensued since perestroika and the move to a market economy is, in some ways, worse."[20] A 1999 Canadian report confirms that depressing estimate: "There is severe pollution of air, water and soil in many regions of the country, caused by spills and routine discharges. This is suspected to have contributed to the recent decline of life expectancy in Russia . . . Industrial pollution affects not only Russia, but also its neighbors. For example, the Nickel metallurgical plant in Norilsk is one of the largest single sources of acid rain on the planet."[21]

It's true that the long-term consequences of this sort of damage can be overstated, which is a danger in itself. Technology refines itself over time, multiplies its own powers, cleans up its own act (when its masters are pushed hard enough). We need not rip the breast of the earth, if clever recycling proves cheaper as well as politically more enticing. We need not deplete unreplaceable resources, if subtle means can be found—as they constantly are, in labs all around the globe—to use materials in smarter ways.[22]

More important, this entire debate will become redundant within a generation or two, like the agonized discussions over how many more horses should be allowed to haul carriages in the thoroughfares of the empire before their dung made the streets entirely disgusting and unpassable . . .

A paradoxically slow rise to the Spike

Estimates of when the Spike is due, and even of its slope and the time remaining before that slope carries us upward faster than we can foresee, remain elusive even to those few informed people who have studied the question. Perhaps that is just as well (although we shall examine some estimated time lines in the final chapter). It would be all too easy to plunge from skepticism to abject worship. That would be a foolish reaction. We will not find ourselves hurled headlong into the Singularity in the next few years, or even the next decade or two. The curve steepens only later, even if that runaway surge is something that many of us might expect to see in our lifetimes.

For now, what is required of us is not reverence but hard thinking and public dialogue. I will adapt the cautious words with which Eric Drexler ended his 1990 afterword to *Engines of Creation*. You might mistakenly imagine that my aim is primarily to promote the technologies leading to the Spike; it is, rather, to promote understanding of those technologies *and their consequences*.

Even so, as Drexler said of his frankly acknowledged worries about the risks and misuse of molecular nanotechnology: "The sooner we start serious development efforts, the longer we will have for serious public debate. Why? Because serious debate will start with those serious efforts, and the sooner we start, the poorer our technology base will be. An early start will thus mean slower progress and hence more time to consider the consequences."

It is a nice paradox. If we postpone our analysis of the path to the Spike, on the understandable grounds that it's too frightening, or too silly, we'll lose the chance to keep our hands on the reins. (And see how the old, passé metaphors remain in charge of our thoughts? Hands on the *reins*?)

Drexler once more: "We need to develop and spread an understanding of the future as a whole, as a system of interlocking dangers and opportunities. This calls for the efforts of many minds." To date, many of the best minds of the human race have been devoted to short-term goals suitable for short-lived people in a volatile, hungry, dangerous world. That will change. Perhaps none of the complex lessons of our long history will have the slightest bearing on our conduct, will offer us any good guidance, in the strange days after the middle of the twenty-first century and beyond. Except, perhaps, as Lem would argue, the austere rules of game theory, and the remote laws of physics.

We will wish to shout against this bleak obliteration of the wisdom of the past. "Love one another!" some will cry. Yes, that is part of our deep grammar, inherited from three million years on the plains of Africa. So too is its hateful complement: "Fear the stranger! Guard the food! Kill the foe!" The Spike might solve these ancient dilemmas by rendering some of them pointless—

why bother to steal another's goods when you can make your own in a matter-compiler?—and others remediable—why shiver in fear of disease and sexism and death, when you can rewrite your DNA codes, bypass mortality, switch gender from the chromosomes up, guided by wonderful augmented minds?

Three versions of the Spike

Or is this, in turn, no better than cargo-cult delusion, the ultimate mistaken reliance for salvation upon some God-in-the-Machine?

Anders Sandberg, one of the few people in the world to think about transhumanist issues in depth, has deplored the way our minds tend to cave in when faced by the possibility of a Spike. The concept is so *immense:* "it fits in too well with our memetic receptors!" Sandberg strove to unpick the varieties of Singularity that have been proposed to date.

One is the Transcension, an approximation to the Parousia where we become more than human, changing by augmentation or uploading into something Completely Different and Unknowable.

A second is the Inflexion Point, the place where the upward scream of the huge sigmoid curve of progress tips over, slowing, and starts to ebb.

The third is the Wall, or Prediction Horizon, the date beyond which we can't predict or understand much of what is going on because it simply gets too weird. While this last version somewhat resembles the first, that is just a side effect of our ignorance. It does not imply, as Transcension can, that with the Spike we enter a realm of spacetime engineering, creation of budded universes, and wholesale rewriting of the laws of physics: what Mitch Porter, borrowing from David Zindell, calls "the Vastening." Vernor Vinge has made it clear that this is the limited kind of singularity he takes seriously.[23]

For all its appeal—precisely *because* of its cryptomystical, pseudoreligious appeal—the Transcension is, Anders Sandberg suggests, the most dangerous of the three versions, since it is the most overwhelming. Many discussions just close with, "But we cannot predict *anything* about the post-singularity world!," ending all further inquiry just as Christians and other religious believers do with, "It is the Will of God." And it is all too easy to give the Transcension eschatological overtones, seeing it as Destiny. This also promotes a feeling of helplessness in many, who see it as all-powerful and inevitable.

Any tendency to worship the Singularity, then, must be resisted firmly. Yet, held in check, subject to scrutiny, might some measure of dread and wonderment in the face of the technologically sublime actually be appropriate? While we look toward the far, far distant closure of the cosmos and the ignition of a homegrown deity in the ashes of ruined stars, perhaps we might use the time—those many, many billions of years—to establish ourselves or our machine offspring among those stars in the aeons before their fall. That is, indeed, a prerequisite of both Dyson's and Tipler's projections, for mind must colonize the universe before it can be transcended.

Spreading humanity to the stars

Luckily enough, a quite precise scheme has been developed to get this program started. Extraordinarily detailed scenarios formulated by Marshall T. Savage and his associates extend from a Millennial Foundation (already in existence) dedicated to dotting the tropical oceans with self-constructing human habitats powered by thermal generators, through to plans for solar-powered bubbles in orbit and domed cities inside lunar craters, to the terraforming of Mars, the building of Dyson clouds of inhabited asteroids girdling the sun in the same orbit as the Earth, and eventually the dissemination of starships throughout the galaxy at close to the speed of light.

We stand, with such an articulated project, at the boundary

between the visionary and the lunatic, the ambitious (surely praiseworthy) and the preposterous (simply laughable). Yet this could be the shape of the upward curve of the Spike, as we attempt to imagine the strictly unimaginable. Start with an oceanic city, Aquarius, built like a castle in clouds but in this case from the very elements suspended in the sea's currents. You sink the taproot of an ocean thermal energy converter (OTEC) platform. Using the 40-degree heat differential between sun-warmed surface waters and the cold drafts from the depths, a turbine can extract as much energy as you'd get by harnessing a hundred-meter waterfall.

Fabricating a floating equatorial city, far from land in international waters, calls for equally lateral but realistic engineering. Make it out of reinforced concrete—not shipped there from factories, but derived directly from the calcium carbonate and trace mineral ions already suspended in the water. These useful ions are drawn to a wire grid by free power from the energy converter. Oddly enough, such *sea-crete* "is stronger and lighter than conventional concrete, and does not lose any of its strength when it dries." Utopias are built from practical if arcane knowledge like this, and not merely from wistful hopes. The Spike will be a human construction before it is a posthuman habitation.

Still, it might seem that a plan for humans to colonize the galaxy over the coming millennium—"in eight easy steps," as Savage puts it cheekily—is as deep into egocentrism as a mind can fall. Perhaps not. Certainly Savage's pragmatic but intoxicating vision is a colossal feat of hubris, but is he therefore (perhaps sharing the victor's belt with Tipler) the new heavyweight champ of triumphalist Faustian science?

Consider the Millennial Foundation's scheme for traversing the galaxy. It is plainly difficult, even using wildly expensive antimatter fuel, to accelerate your starship to nearly the speed of light. Yet anything markedly slower takes the joy out of the trip. You'll start it, but your great-great-grandchildren will be the ones to land on some new world. Even with nanite longevity or journey-time passed as a VR upload, you could get rather bored. Here's a transhuman alternative: build yourself an electromagnetic catapult to throw your sleek starship across deep space at 99 percent of light-speed. The drawbacks are nothing more formidable, after all, than mere engineering obstacles. The launcher, an extended linear accelerator, will need to be ... *450 thousand million kilometers long* ... That's fifty times the extent of the entire solar system. A hundred trillion tons of mass. But we can *do that*! Just dismantle an asteroid or two. The nanites will perform the hard work.

Build a pair of these star bridges (one in your home system, the other at the far end, so you'll first need to send a Santa Claus machine the slow way, pushed by laser beams or dragging its fuel with it), and you can throw and catch pods fantastically fast at zero overall energy cost: what's stored in velocity on the outward voyage is recovered by slowing the pod at its destination. Of course, you'll still need to shield your starships, which must plow through thin but lethal interstellar gunge at nearly three hundred thousand kilometers a second, churning out X rays and other nasty stuff as the dust particles impact the eroding nose of the craft ...

Savage claims that his itemized—some would say vainglorious—project to enliven a dead cosmos with loving human minds and bodies is "manifest destiny." But it is a politically okay green destiny, even if it does happily predict populations that soar until a rebuilt solar system contains 50 million transcendent posthuman swarm entities, each comprising 100 billion people. The Spike! It is an epic vision, perhaps the most grandiose but detailed in the history of political prophecy. Just possibly, the third millennium will be remembered for the realization of this kind of dream, long after everything else is gone into oblivion or singularity.

The Spike is us

Already, even before the silver cities in the sky are sent aloft on their rainbow laser beams, we can look outward upon the uni-

verse with the borrowed eyes of earth- and space-based obser-
vatories, and with the great corrected mirrors of the orbiting
space telescope. On 14 February 1990, as the late Carl Sagan
reminded us, the *Voyager* space probe turned its electronic gaze
back at Earth. It was already beyond the orbit of Neptune, currently the farthest world from the Sun. This creation of human
hands and minds rushed on into deep space at 65,000 kilometers
an hour. It was six billion kilometers from home, sending back
family snaps that took five and a half hours to reach us. Our blue
world's portrait was captured in the midst of an image composed
of 640,000 pixels—little individual dots, mostly black. Earth was
one pixel, a single point of light. James Elroy Flecker's elegiac
poem "The Golden Journey to Samarkand" foresaw this moment:

> When the great markets by the sea shut fast
> All that calm Sunday that goes on and on:
> When even lovers find their peace at last,
> And Earth is but a star, that once had shone.

It is a perspective to inflame the heart with a sense of our true
place in the universe. We are not at its center, not even very
special, born from random mutations and ruthless selection. Yet
we can send our emissaries into the cosmos, driven by nothing
more potent than a dream—and, of course, a vast and growing
corpus of knowledge about how that universe actually works.
How extraordinary, how chilling, for us planet-bound creatures
to gaze, as we may in this wondrous new century, at a full-color
portrait of the pitted asteroid Ida and its tiny moon, at Australia's
red-brown bulge amid blue ocean and white cloud under the
hanging Hubble telescope, at the islands, the archipelagos, the
filaments of stars stretching into the occlusion of space and time.
Yes, science is made by humans, and its knowledge is contaminated by our local limitations. But we cup our hands, and the
cosmos fills them to overflowing.

It is no mean ambition to understand this: that in a world
born out of nothingness, evolved from noise by mutation and

brute survival, undesigned and meaningless except for whatever meaning we choose to introduce recursively in our evolved minds and hearts, passion and mind might end by suffusing the cosmos and transforming it utterly.

Summary: Paths and Time-lines to the Spike

[R]APID AND ACCELERATING ADVANCES IN SCIENCE AND TECHNOL-
OGY—AND THE SOCIAL, POLITICAL, AND PSYCHOLOGICAL CHANGES
THAT FOLLOW AND COMPLEMENT THEM—ARE TRANSFIGURING SOCI-
ETY, ECONOMY, POLITICS, AND WARFARE IN PROFOUND WAYS. IN ITS
SIMPLEST AND MOST FUNDAMENTAL SENSE, THIS TRANSFORMATION IS
A SHIFT FROM AN INDUSTRIAL AGE TO A KNOWLEDGE ERA . . . THE RISE
OF THE INDUSTRIAL ERA KINDLED SOCIAL AND ECONOMIC INFERNOS,
FROM THE AWFUL CONDITIONS FACED BY MILLIONS OF FACTORY
WORKERS TO DEVASTATING SOCIAL MOVEMENTS . . . [W]E ARE UN-
DERGOING PRECISELY SUCH A TRANSFORMATION . . . HISTORICAL
TRANSITIONS CAN EXACT A SEVERE HUMAN TOLL.
  —MICHAEL J. MAZARR, GLOBAL TRENDS 2005 1999, P. 2

I wish I could show you the real future, in detail, just the way
it's going to unfold. In fact, I wish I knew its shape myself. But
as you read the earlier chapters, you'll have encountered one repeated mantra:

"The unreliability of trends is due precisely to *relentless, unpredictable change,* which makes the future interesting but also
renders it opaque."

I stated that right up front, in chapter 2, and I still stand by
it. Recall Stanislaw Lem's daunting, exhilarating insight: "It is a
law of civilizational dynamics that instrumental phenomena grow
at an exponential rate . . . the existence of future generations totally transformed from ours would remain an incomprehensible
puzzle for us, even if we could express it."

On the other hand—

Is that strictly true? There are some negative constraints we
can feel fairly confident about. We *can* be fairly sure that Ptole-
maic astronomy is not due for a surprise revival (despite the

mind-numbing persistence of astrology, a doctrine based on a model of the heavens known for the last four hundred years to be untrue). The sheer reliability and practical effectiveness of quantum theory, and the robust way relativity holds up under strenuous challenge, argues that both will remain at the core of future science—in *some* form, which is rather baffling, since at the deepest levels they disagree with each other about what kind of cosmos we inhabit.[1] In other words, we do already know a great deal, a tremendous amount, corroborated knowledge will not go away. But we also know for certain that our jigsaw picture of the world is made up of large chunks with jagged edges that don't always meet. Maybe we're mixing up several quite different jigsaws. Maybe we just need to find a cleverer way to slide the pieces past each other and into the big frame.

Meanwhile, the Spike apparently looms ahead of us: a horizon of ever-swifter change we can't yet see past. We can try to imagine the unimaginable, though, up to a point. That is what scientists and artists (and visionaries and explorers) have always attempted, Lem among them, as part of their job description. So let's see if we can draw together the threads examined in this book, sketch a number of possible pathways into and beyond the Singularity.

First, though, one must ask if that outcome is even remotely desirable. In mid-March 2000, the chief scientist and cofounder of Sun Microsystems, Bill Joy, published a now much-discussed warning that took such prospects very seriously indeed. And dreaded them. He declared with trepidation: "The vision of near immortality that [Ray] Kurzweil sees in his robot dreams drives us forward; genetic engineering may soon provide treatments, if not outright cures, for most diseases; and nanotechnology and nanomedicine can address yet more ills. Together they could significantly extend our average life span and improve the quality of our lives. Yet, with each of these technologies, a sequence of small, individually sensible advances leads to an accumulation of great power and, concomitantly, great danger" (*Wired* magazine, April 2000).[2] He is right to be concerned, but I believe the risks are worth taking. Let's consider the way this deck of novelties might play out.

We need to simplify in order to do that. That is, we need to take just one thread at a time and give it priority, treat it as if it were the only big change coming down the pike, or at least the main one, modulating everything else that falls under its shadow. It's a risky gambit, since it has never been true in the past and will not strictly be true in the future. The only exception is the dire (and possibly false) prediction that something we do, or something from beyond our control, brings down the curtain, blows the whistle to end the game. So let's call that option

[A i] No Spike, because the sky is falling

In the second half of the twentieth century, people feared that nuclear war (especially nuclear winter) might snuff us all out. Later, with the arrival of subtle sensors and global meteorological studies, we worried about ozone holes and industrial pollution and a human-induced greenhouse effect combining to blight the biosphere. That's still an urgent concern. Later still, the public became aware of the small but assured probability that sooner or later our world will be struck by a "dinosaur-killer" asteroid, which could arrive at any moment. For the longer term still, we started to grasp the cosmic reality of the sun's mortality, and hence our planet's: solar thermodynamics will brighten the sun in the next half billion years, roasting the surface of our fair world and killing everything that still lives upon it. Beyond that, the universe as a whole will surely perish one way or another, in heat or cold, unless vast post-Spike mentalities take its fate into Their Hands and rescind it.

Of course, start thinking seriously about posthumans on the far side of the Spike and you have to wonder where their extraterrestrial equivalents are. Maybe they already suffuse the galaxy, as humankind and our descendants surely will do in another million or two years if there are no ETs out there in prior possession of the real estate.

Those are ways in which we have imagined the *end* of life on Earth, or at least the end of humanity. Take a more optimistic view of things. Suppose we survive as a species, and maybe as individuals, at least for the medium term (forget the asteroids and *Independence Day*). That still doesn't mean there must be a Spike, at least in the next century or two. Perhaps artificial intelligence will be far more intractable than Moravec, Kurzweil, and other enthusiasts proclaim. Perhaps molecular nanotechnology stalls at the level of MEMS, microscale machines that have an impact but never approach the fecund cornucopia of a true mint. Or perhaps matter compilers will be developed, but the security states of the world agree to suppress them, imprison or kill their inventors, prohibit their use at the cost of extreme penalties. Then we have option

[A ii] No Spike, steady as she goes

This obviously forks into a variety of alternative future histories, the two most apparent being

[A ii a] Nothing much ever changes ever again

which is the day-to-day working assumption I suspect most of us default to, unless we force ourselves to think hard. It's like that illusion of unaltered identity that preserves us sanely from year to year, decade to decade, allows us to retain our equilibrium in a lifetime of such smashing disruption that some people still living went through the whole mind-wrenching transition from agricultural to heavy industrial to knowledge/electronic societies. Such simple continuity is an illusion, and perhaps a comforting one, but I think we can be pretty sure the future is not going to stop happening just because we've arrived in the twenty-first century.

The clearest alternative to that impossibility is

[A ii b] Things change slowly (haven't they always?)

Well, no, they haven't. This option does pretend to acknowledge our previous century's vast changes, but insists that, even so, *human nature itself* has not changed. True, the argument admits, racism and homophobia are increasingly despised rather than mandatory. True, warfare is now widely deplored (at least in rich, complacent places) rather than extolled as honorable and glorious. Granted, people who drive fairly safe cars while chatting on the mobile phone live rather . . . strange . . . lives, by the standards of the horse-and-buggy era only a century behind us. Still, once everyone in the world is drawn into the global market, once peasants in India and villagers in Africa also have mobile phones and learn to use the Internet and buy from IKEA, things will . . . *settle down*. Nations overburdened by gasping population pressures will pass through the demographic transition, easily or cruelly, and we'll top out at around 10 billion humans living a modest but comfortable, ecologically sustainable existence for the rest of time (or until that big rock arrives).

A bolder variant of this model is

[A iii] Increasing computer power will lead to near-human-scale AI, and then stall

But why should technology abruptly falter in this fashion? Perhaps there is some technical barrier to improved miniaturization, or connectivity, or dense, elegant coding (but experts argue that there will be ways around such roadblocks, and advanced research already points to some likely prospects: quantum computing, nanoscale processors). On the other hand, natural selection has not managed to leap to a superintelligent variant of humankind in the last hundred thousand years, which implies that mutations in that direction are not common, or have associated deficits (as Eliezer Yudkowsky believes true of his own gifts). Whatever small genetic changes permitted protohumans to diverge from chimpanzees over a million years ago (changes supporting sign and spoken language, say, and hence self-reinforcing culture), it looks as if no equivalent small heritable modification will build superman. If so, maybe there is some structural reason why brains top out at the Murasaki, Einstein, or van Gogh level. (But why should such limits apply to machines with quite different architecture?)

AI research might reach the low-hanging fruit, all the way to human equivalence, and then find it impossible (even with machine aid) to find a path through murky search space to a higher level of mental complexity. Still, using the machines we already have will not leave our world unchanged. Far from it. And even if this story has some likelihood, a grislier variant seems even more plausible.

[A iv] Things go to hell, and if we don't die we'll wish we had

This isn't the nuclear-winter scenario, or any other kind of doom by weapons of mass destruction—let alone gray nano goo, which by hypothesis never gets invented in this denuded future. It's a perspective caught vividly in this quietly despairing statement by my friend Bruce Gillespie:

> the essential fact of the 20th century was the unending, ferocious assault by the human species upon every other species on the planet, and the pushing of all life systems to their limit. The human species is basically a nasty bug that has infected the planet, which must be either got rid of or contained. It only needs the slippage of one major life system, such as (for instance) the destruction of life in the top centimeter of sea water and everything you're talking about will have no point whatsoever. If the human species cannot learn to be one species among millions, and get back within limits, then there's not much hope of reaching 2050, let alone 3000. So-called "Western civilization" is a luxury the world cannot afford: it's basically a process for creating ever-increasing piles of shit (and increasing at an ever faster rate) in which to drown, while grabbing everything from the rest of the world in order to add to the shit pile. All the damage has already been done; it's just a matter of waiting to see the interesting/horrifying ways in which the shit hits the fan.

That assessment will have Green readers nodding and extropians frothing at the mouth. There's something to be said for its bleak vision, but the key lament falls apart the moment you realize that a lot of the "shit" Western culture piles up is CDs of Beethoven and Bach and the Beastie Boys, the hygiene and medical remedies that save us all a lot of pain, the instrumentalities of art and science and architecture and good cooking that inspire our lives and offer us fun as well as relief from humankind's heritage of face-grinding toil. Besides, that melancholy diagnosis of a world deluged by trash is a kind of optical illusion, projecting a refuse pile in a Philippines slum over the clean, comfortable environments most of us inhabit fairly happily.

Still, those benefits demand a toll from the planet's resource base, and our polluted environment. The rich nations, numerically in a minority, notoriously use more energy, materials than the rest, and plainly pour more crap into air and sea. That can change—*must* change, or we are all in bad trouble—but in the short term one can envisage a nightmare decade or two during which the Third World "catches up" with the wealthy consumers, burning cheap, hideously polluting soft coal, running the exhaust of a billion and more extra cars into the biosphere... Some Green activists mock "technical fixes" for these problems, but those seem to me our last best hope.[3] We are moving toward manufacturing and control systems very different from the wasteful, heavy-industrial, pollutive kind that helped drive up the world's surface temperature by 0.4 to 0.8 degrees Celsius (0.7–1.4 degrees Fahrenheit) in the twentieth century.[4]

Pollsters have noted with incredulity during the last decade that people overwhelmingly state that their individual lives are quite contented and their prospects good, while agreeing that the nation or the world generally is headed for hell in a handbasket. It's as if we've forgotten that the nonstop vice and brutality of television entertainments, and even news broadcasts, do *not* reflect the true state of the world. Almost the reverse is true: we revel in such violent cartoons because, for almost all of us, our lives are comparatively placid, safe and measured. If you doubt this, go back in time and live for a while in medieval Paris, or Paleolithic Egypt (you're not allowed to be a noble).

Roads from here and now to the Spike

I assert that all of these *No Spike* options are of low probability, unless they are brought forcibly into reality by the hand of some Luddite demagogue using our confusions and fears against our own best hopes for local and global prosperity. If I'm right, we are then pretty much on course for an inevitable Spike. We might still ask: what, exactly, is the motor that will propel technological culture up its exponential curve?

Here are seven distinct candidates for paths to the Spike (separate lines of development that in reality will interact, generally hastening, sometimes slowing each other):

[B i] Increasing computer power will lead to human-scale AI, and then swiftly self-bootstrap to incomprehensible superintelligence

This is the "classic" model of the singularity, the path to the ultraintelligent machine and beyond allegorized by Stanislaw Lem in his short fiction "Golem XIV." This remarkable superintelligent computer was the latest in a series of self-augmenting machines that variously went mad, fell silent, or vanished from the sight of their makers.[5] Lem proposed a "toposophy," somewhat like the underpinnings to Vinge's fictional galactic Zones, a kind of topology of thought. It is reasonable that minds better than our own will function in ways we unmodified *Homo sapiens* just can't grasp, certainly can't emulate or equal. Little wonder, then, that the exact nature of a planet containing one or more of these new intelligences, let alone thousands or millions, must remain opaque to our early twenty-first-century gaze. If this path is chosen deliberately, rather than by simple accidental convergence, we are in Yudkowsky's future, which is examined in more detail below, as option [C].

Suppose there is no abrupt leap from today's moderately fast machines to a fully functioning artificial mind equal to our own, let alone its self-redesigned kin. If we can trust Moore's law as a guide (and strictly, as I have reiterated, we can't, since it's only a record of the past rather than an oracle), we get the kinds of time lines presented by Ray Kurzweil, Hans Moravec, Michio Kaku, Peter Cochrane, and others. Let's briefly recall and sample those predictions:

- Cochrane: the British Telecom futures team, led by their guru Peter Cochrane, saw human-level machines as early as 2016. Their remit did not encompass a sufficiently deep range to sight the singularity.
- Kurzweil: around 2019, a standard cheap computer has the capacity of a human brain, and some claim to have met the Turing test (that is, passed as conscious, fully responsive minds). By 2029, such machines are a thousand times more powerful. Machines not only ace the Turing test, they *claim to be conscious,* and are accepted as such. His sketch of 2099 is effectively a Spike: fusion between human and machine, uploads more numerous than the embodied, immortality.
- Merkle: while Ralph Merkle's special field is nanotech, this plainly has a possible bearing on AI. His is the standard case, although the time line is still "fuzzy":[6] various computing parameters go about as small as we can imagine between 2010 and 2020, if Moore's law holds up. To get there will require "a manufacturing technology that can arrange individual atoms in the precise structures required for molecular logic elements, connect those logic elements in the complex patterns required for a computer, and do so inexpensively for billions of billions of gates." So the imperatives of the computer hardware industry will create nanoassemblers by 2020 at latest. Choose your own timetable for the resulting Spike once both nano and AI are in hand. Merkle still cites the August 1995 *Wired* poll of experts (chemistry professor Robert Birge, materials science professor Donald W. Brenner, Drexler, computer scientist J. Storrs Hall [JoSH], and Nobelist chemist Richard E. Smalley) on several aspects of a time line to working nano. Here are their now somewhat superannuated but interesting estimates:

	Birge	Brenner	Drexler	Hall	Smalley
Molecular Assembler:	2005	2025	2015	2010	2000
Nanocomputer:	2040	2040	2017	2010	2100
Cell Repair:	2030	2035	2018	2050	2010
Commercial Product:	2002	2000	2015	2005	2000

- Moravec: multipurpose "universal" robots by 2010, with "humanlike competence" in cheap computers by around 2039—a more conservative estimate than Ray Kurzweil's, but astonishing nonetheless. Even so, he considers a Vingean singularity as likely within fifty years.
- Kaku: no computer expert, superstring physicist Michio Kaku surveyed some 150 scientists in devising a profile of the next century and farther. He concludes broadly that from "2020 to 2050, the world of computers may well be dominated by invisible, networked computers which have the power of artificial intelligence: reason, speech recognition, even common sense."[7] In the next century or two, he expects humanity to achieve a Type I Kardeshev civilization, with planetary governance and technology able to control weather but essentially restricted to Earth. Only later, between 800 and 2500 years farther on, will humanity pass to Type II, with command of the entire solar system. Once the consensus dream of science fiction, this must now be seen as excessively conservative.
- Vinge: as we noted at the outset, Vernor Vinge's part-playful, part-serious proposal that a singularity was imminent puts the date at around 2020, marking the end of the human era. Maybe as soon as 2014.
- Yudkowsky: once we have a human-level AI able to understand and redesign its own architecture, there will be a swift escalation into a Spike. Could be as soon as 2010, with 2005 and 2020 as the outer limits, if the Singularity Institute has anything to do with it (see option [C], below).

*[B ii] Increasing computer power will lead to direct augmentation of
human intelligence and other abilities*

Why build an artificial brain when we each have one already?
Well, it is regarded as impolite to delve intrusively into a living
brain purely for experimental purposes, whether by drugs or surgery (sometimes dubbed "neurohacking"), except if no other
course of treatment for an illness is available. Increasingly subtle
scanning machines are now available, allowing us to watch as the
human brain does its stuff, and brave pioneers are coupling chips
to parts of themselves, but few expect us to wire ourselves to
machines in the immediate future. That might be mistaken, however. Professor Kevin Warwick, of Reading University, successfully
implanted a sensor-trackable chip into his arm in 1998. A year
later, he allowed an implanted chip to monitor his neural and
muscular patterns, then had a computer use this information to
copy the signals back to his body and cause his limbs to move;
he was thus a kind of puppet, driven by the computer signals.
He plans experiments in which the computer, via similar chips,
takes control of his emotions as well as his actions.[8]

As we gradually learn to read the language of the brain's neural
nets more closely, and finally to write directly back to them, we
will find ways to expand our senses, directly experience distant
sensors and robot bodies (perhaps giving us access to horribly
inhospitable environments like the depths of the oceans or the
blazing surface of Venus). Instead of hammering keyboards or
calculators, we might access chips or the global net directly via
implanted interfaces. One risk with this proposed shortcut to
augmented intelligence is infection. Cutting open the skin, especially with the goal of leaving something alien inside the body,
is a major hazard.

Perhaps, instead, sensitive external monitors will track brain
waves, myoelectricity (muscles), and other indices, and even impose patterns on our neurons using powerful, directed magnetic
fields. Augmentations of this kind, albeit rudimentary, are already
seen at the lab level. Perhaps by 2020 we'll see boosted humans

able to share their thoughts directly with computers. If so, it is a fair bet that neuroscience and computer science will combine to map the processes and algorithms of the naturally evolved brain, and try to emulate it in machines. Unless there actually *is* a mysterious nonreplicable spiritual component, a soul, we'd then expect to see a rapid transition to self-augmenting machines—and we'd be back to path Bi.

[B iii] Increasing computer power and advances in neuroscience will lead to rapid uploading of human minds

On the other hand, if Bii turns out to be easier than Bi, we would open the door to rapid uploading technologies. Once the brain/mind can be put into a parallel circuit with a machine as complex as a human cortex (available, as we've seen, somewhere between 2020 and 2040), we might expect a complete, real-time emulation of the scanned brain to be run inside the machine that has copied it. Again, unless the "soul" fails to port over along with the information and topological structure, you'd then find your perfect twin (although grievously short on, ahem, a body) dwelling inside the device.

Your uploaded double would need to be provided with adequate sensors (possibly *enhanced,* compared with our limited eyes and ears and tastebuds), plus means of acting in the world with ordinary intuitive grace (via physical effecters of some kind—robotic limbs, say, or a Moravecian bush robot telepresence). Or perhaps your upload twin would inhabit a cyberspace reality, less detailed than ours but more conducive to being rewritten closer to heart's desire. How soon is such VR likely? A rather cautious expert, MIT's Michael Dertouzos, considers "computationally intense simulations, such as full-immersion virtual reality with haptic [touch-mimicking] suits, goggles, and trackable helmets . . . will take one or two decades before they are possible at a moderate quality level and more time than that before they're affordable."[9] Such VR protocols should lend themselves readily to life as an uploaded personality.

Once mind uploading is shown to be possible and tolerable

or, better still, enjoyable, we can expect at least some people to copy themselves into cyberspace: to "homestead the noösphere," if I may bend Eric Raymond's phrase. How rapidly this new world is colonized will depend on how expensive it is to port somebody there, and to sustain them. Computer storage and run-time should be far cheaper by then, of course, but still not entirely free. As Robin Hanson has argued, the problem is amenable to traditional economic analysis. "I see very little chance that cheap fast upload copying technology would not be used to cheaply create so many copies that the typical copy would have an income near 'subsistence' level."[10] On the other hand, "If you so choose to limit your copying, you might turn an initial nest egg into fabulous wealth, making your few descendants very rich and able to afford lots of memory."

If an explosion of uploads is due to occur quite quickly after the technology emerges, early adopters would gobble up most of the available computing resources. But this assumes that uploaded personalities would retain the same apparent continuity we fleshly humans prize. Being binary code, after all (however complicated), such people might find it easier to alter them-selves—to rewrite their source code, so to speak, and to link themselves directly to other uploaded people, and AIs if there are any around. This looks like a recipe for a Spike to me. How soon? It depends. If true AI-level machines are needed, and perhaps medical nanotechnology to perform neuron-by-neuron, synapse-by-synapse brain scanning, we'll wait until both technologies are out of beta-testing and fairly stable. That would be 2040 or 2050, I'd guesstimate.

[B iv] Increasing connectivity of the Internet will allow individuals or small groups to amplify the effectiveness of their conjoined intelligence, leading swiftly to an emergent AI

The most extreme version of this narrative is Dan Clemmensen's path to the Spike, explored in fiction in John Barnes's *Mother of Storms*. In essence: routine disseminated software advances will create (or evolve) ever smarter and more useful support systems

for thinking, gathering data, writing new programs—and the outcome will be a "in-one-bound-Jack-was-free" surge into AI. That
is the garage band model of a singularity, and while it has a
certain cheesy appeal, I very much doubt that's how it will happen. If it does, it will surely not occur as soon as 2006, Clemmensen's bravely posted bet.

But the Internet is growing and complexifying at a tremendous
rate. It is barely possible that one day, as Arthur C. Clarke suggested decades ago of the telephone system, it will just . . . *wake
up*. After all, that's what happened to a smart African ape, and it
and its close genetic cousins weren't already designed to handle
language and mathematics. (On the other, very important hand,
apes *were* selected to handle copious *meaningful* inputs from a
world they had to survive in by their wits, which is not true of
today's computer programs.)

Consider Robert J. Bradbury's projection of significant, cheap,
and fast nano development, a parallel track to home-grown computer sophistication and feeding back into it:

> The technologies you need to construct assemblers are going
> to become increasingly commonplace. If trends in the min-
> iaturization of disk drives continue, I expect in a few years
> people could disassemble a couple of disk drives and have
> much of the hardware required for precise 2-D positioning.
> You can generate buckytube AFM tips in an electric arc de-
> vice with carbon electrodes. Any hobbyist can now purchase
> piezoelectric positioners, lasers for measuring tip position,
> etc. In short it will be extremely hard to control access to
> the materials you need to produce AFMs; from there the
> path to assemblers is only a matter of creativity and time.
> An alternate path would use protein synthesizers available
> in most university molecular biology departments. Do your
> design on a computer, synthesize the proteins, let them self-
> assemble and voila . . . [11]

And Vernor Vinge's own most recent picture of a plausible 2020,
cautiously *sans* singularity, emphasizes the role of embedded

computer networks so ubiquitous that finally they link into a kind of cyberspace Gaia, even merge with the original Gaia, that geological and biological macroecosystem of the globe.[12] With evermore embedded computers inside every gadget and item of clothing, providing users visual overlays to create "consensual imagery" and instant downloaded "remote helping" from distant collaborators, embedded networks would "spread beneath the Net, supporting it much as plankton supports the ocean ecology." Perhaps the software running on this vast disseminated system, "*grown* and *trained*, rather than written," might indeed wake up one day. It's alive!

[B v] Research and development of microelectromechanical systems (MEMS) and fullerene-based devices will lead to industrial nanoassembly, and thence to "anything boxes"

Here we have the "classic" molecular nanotechnology pathway, as predicted by Drexler's Foresight Institute and NASA Ames,[13] and also by the mainstream of conservative chemists and adjacent scientists working in MEMS, and funded nanotechnology labs around the world. In a 1995 *Wired* article, Eric Drexler predicted nanotechnology within twenty years. Is 2015 too soon? Not, surely, for the early-stage devices under development by Zyvex, who hope to have at least preliminary results by 2010, if not sooner. For many years AI was granted huge amounts of research funding, without much result (until recently, with a shift in direction and the wind of Moore's law at its back). Nano is now starting to catch the research dollars, with substantial investment from governments (the United States, Japan, even Australia) and megacompanies such as IBM. The prospect of successful nanotech is exciting, but should also make you afraid, very afraid.

Thomas McCarthy's analysis of the risks in an unstable world, discussed in chapter 7, is all too convincing. If nano remains (or rather, becomes) a closely guarded national secret, contained by munitions laws, a new balance of terror might take us back to something like the Cold War in international relations—but this would be a polyvalent, fragmented, perhaps tribalized balance.

Or building and using nanotech might be like the manufacture of dangerous drugs, toxins, or nuclear materials: centrally produced by big corporations' mints, under stringent protocols (you hope, fearful visions of Homer Simpson dancing in the back of your brain), except for those in Colombia and the local bikers' fortress . . .

Or it might be like the Ma & Pa business in *Unbounding the Future:* a local plant equal, perhaps, to a used-car yard, with a fair-sized goop pool, mass transport shifting raw or partly processed feedstocks in and finished product out. This level of implementation might resemble a small Internet server, with some hundreds or thousands of customers. One might expect the technology to grow more sophisticated quite quickly, as minting allows the emergence of cheap and amazingly powerful computers. Ultimately, we might find ourselves with the fabled anything box in every household, protected against malign uses by an internal AI system as smart as a human, but without human consciousness and distractibility. We should be so lucky. But it could happen that way.

A quite different outcome is foreshadowed in a prescient 1959 novel by Damon Knight, *A for Anything,* in which a "matter duplicator" leads not to utopian prosperity for all but to cruel feudalism, a regression to brutal personal power held by those clever thugs who manage to monopolize the device. A slightly less dystopian future is portrayed in Neal Stephenson's satirical but seriously intended *The Diamond Age,* where tribes and nations and new optional tetherings of people under flags of affinity or convenience tussle for advantage in a world where the basic needs of the many poor are provided free, but with galling drab uniformity, at large-scale street-corner matter compilers owned by authorities. That is one way to prevent global ruination at the hands of crackers, lunatics, and criminals, but it's not one that especially appeals—if an alternative can be found.

Meanwhile, will nanoassembly allow the rich to get richer—to hug this magic cornucopia to their selfish breasts—while the poor get poorer? Why should it be so? Even in a world of 10 billion flesh-and-blood humans (ignoring the uploads for now), there's plenty of space for everyone to own decent housing, transport, clothing, arts, music, sporting opportunities . . . once we grant the ready availability of nano mints. This issue has been analyzed to surprising effect by Robert J. Bradbury.[14] Over the last 10,000 years, he notes, human activity has added 185 petagrams of carbon to the atmosphere, about 31,000 kilos for each person now alive. Using assemblers in a fair, ecologically responsible fashion, what can you get by extracting that excess carbon and reusing it to build a house and other consumer desirables? And how long would it take to compile the goodies?

Start by buying between two and eight acres of cheap land, depending on your latitude and cloud cover. Your nano compiler will grow you solar cells that cover most of the land, providing 400,000 watts a day, powering the compilation of around 10 kilos of materials per hour. (Calculations by Robert Freitas, who allows only 100 kilowatts per person—stringently restricting the total mass and energy of nanoconstructors—would thus imply an average of four people per large house.)[15] Feedstocks from air and soil are almost free. Make your daily food in the first quarter hour. Very conservatively, a 2600-square-foot house (34,000 kilos) will take five months to grow and assemble. Build a huge swimming pool as part of the assembly, as a by-product of mining aluminum for sapphire modules and maybe as a heat sink for another big project: since there's a lot of silicon in the rock along with that aluminum, build a million-kilo basement computer to run the place and handle your eventual upload. Its waste heat will warm your pool. A Bill Gates mansion, Bradbury estimates, would take a little over eight years to build. Not an overnight miracle, but then again it took Gates more than eight years' work, starting from scratch, to get wealthy enough to build his. When can we expect this? With open-source and other nonprofit development of the software, "between 2020 and 2030."

Sounds mad. Actually, friendly critics swiftly pointed out that Bradbury's 1999 estimates might be too cautious by a factor of 100. If so, you could build your mansion in three weeks, at the cost of chewing up more power, and spend some of the saved time compiling the yacht (or maybe the one-stage surface-to-space diamond rocket). But that way you'd need to *buy* the rarer feedstock components from a mining company or utility, an option that still might be quite inexpensive in today's terms. Is there an additional cost in global heat pollution? Robert Freitas examined the "hypsithermal limit" (planetary heat tolerance), deriving daily safe limits of between 100 and 1000 kilowatts per person. Bradbury, recall, opted for 400 KW. But with all that excess carbon being drained from the atmosphere, putting the greenhouse effect conveniently into reverse, we might need all the heat we can get. Entire international lawmaking empires will arise to adjudicate these matters, no doubt, giving many displaced lawyers something to do (until they are replaced by smart AI agents and legal-expert systems).

Why would the rich permit the poor to own the machineries of freedom from want? Some adduce benevolence, others prudence. Above all, perhaps, is the basic law of an information/knowledge economy: the more people who are thinking and solving and inventing and finding the bugs and figuring out the patches, the better a nano minting world is for everyone (just as it is for an open source AI computing world). Besides, how could they stop us?[16] (Well, by brute force, or in the name of all that's decent, or for our own moral good. None of these methods will long prevail in a world of free-flowing information and cheap material assembly. Even China has trouble keeping dissidents and mystics silenced.)

The big necessary step is the prior development of early nano assemblers, and we have seen that this will be funded by university and corporate (and military) money for researchers, as well as by increasing numbers of private investors who see the marginal payoffs in owning a piece of each consecutive improvement in micro- and nanoscale devices. So yes, the rich will get richer— but the poor will get richer too, as by and large they do now, in the developed world at least. Not *as* rich, of course, nor *as* fast. By the time the nano and AI revolutions have attained maturity, these classifications will have shifted ground. Economists insist that rich and poor will still be with us, but the metric will have changed so drastically, so strangely, that we here-and-now can make little sense of it. Anders Sandberg put it thus:

> The solar system might be colonised by a huge explosion of nanotech conversion, computronium-building and upload-copying, but once this has reached its inflexion point other activities will likely take over. This explosion would at the same time be competitive for all the resources, and likely post-economical in the sense that the huge amount of resources becoming accessible would likely make each individual richer and richer unless they multiplied faster than the econosystem could grow. Now, I'm of the opinion that . . . it doesn't make much sense to endlessly copy one-self for the colonization effort—most of it will be easily managed routine, requiring relatively few uploads. Where a lot of uploads would be necessary is the complex, growing areas of human activities (such as lawyers and consultants). But here the economics doesn't have to grow as fast, it is really a balance between values and economics—it could explode (there are so many lawyers that everybody needs one—even the lawyers themselves) but it could also (and I believe this is more likely) diversify into an evolutionary radiation into the new resources, where the distribution of resources would be less competitive.[17]

Some readers will find this kind of projection simply preposterous, especially its maddeningly calm tone. I agree that it's hard to take post-Spike scenarios like this absolutely seriously—but not because they are foolish or self-indulgent. On the contrary: Sandberg's analysis is just what you get when you accept the singularity premise literally, and push hard on it. That is where we are heading. Expect some interesting court cases as early Exes head out into the solar system and start to dismantle asteroids and even the odd moon or planet.

[B vi] Research and development in genomics (the Human Genome Project, Celera, etc.) will lead to new "wet" biotechnology, lifespan extension, and ultimately to transhuman enhancements.

This is a quite different but parallel approach, and increasingly I see experts arguing that it is the shortcut to mastery of the worlds of the very small and the very complex. Biology, not computing! Wet and soft, not hard and dry! After all, bacteria, ribosomes, viruses, cells for that matter, are already operating beautifully at the micro- and even the nanoscales.

Still, even if technology takes a major turn away from mechanosynthesis and "hard" minting, this approach will require a vast armory of traditional and innovative computers and appropriately ingenious software. The IBM project Blue Gene is a huge system of parallel processors designed to explore protein folding, crucial now that the genome projects have compiled their immense catalog of genes. Knowing a gene's recipe is of little value unless you know, as well, how the protein it encodes twists and curls in three-dimensional space. That is the promise of the first couple of decades of the twenty-first century, and it will surely unlock many secrets and open new pathways, perhaps by learning directly from nature with "biomimetic machines."[18]

Exploring those paths will need all the help the molecular biologists can get from advanced computers, virtual reality displays, and AI adjuncts. And, once again, we can reasonably expect those paths to track right into the foothills of the Spike. Put a date on it? Nobody knows—but recall that DNA was first decoded in 1953, and by around half a century later the whole genome is in the bag. How long until the next transcendent step—complete understanding of all our genes, how they express themselves in tissues and organs and abilities and behavioral bents, how they can be tweaked to improve them dramatically? Cautiously, the same interval: around 2050. More likely (if Moore's law keeps chugging along), half that time: 2025 or 2030.

The usual timetable for the Spike, in other words.

[C] The Singularity happens when we go out and make *it happen.*

That's Yudkowsky's sprightly, in-your-face declaration of intent, which dismisses as jejune and uncomprehending all the querulous cautions about the transition to superintelligence and the Singularity on its far side.

In essence, this analysis claims, just attaining human-level AI is more than enough for the final push to a Spike. How so? Don't we need unique competencies to do that? Isn't the emergence of ultraintelligence, either augmented-human or artificial, the very definition of a Vingean singularity?

Yes, but this is most likely to happen when a system with the innate ability to view and reorganize its own cognitive structure approaches the conscious power of a human brain. A machine might have that facility, since its programming is listable: you could literally print it out—in many, many volumes—and check each line. Not so an equivalent human, with our protein spaghetti brains, compiled by gene recipes and chemical gradients rather than exact algorithms; we clearly just can't do that.

So you get intelligent design turned back upon itself, a cascading multiplier that has no obvious bounds. The primary challenge becomes software rather than hardware. The raw petaflop hardware end of the project is chugging along nicely now, mapped by Moore's law, but even if it tops out, it doesn't matter. A self-improving seed AI could run glacially slowly on a limited machine substrate. The point is, so long as it has the capacity to improve itself, at some point it will do so convulsively, bursting through any architectural bottlenecks to *design* its own improved hardware, maybe even build it (if it's allowed control of tools in a fabrication plant). So what determines the arrival of the Singularity is just the amount of effort invested in getting the original seed software written and debugged.

This particular argument is detailed in Yudkowsky's ambitious Web documents "Coding a Transhuman AI," "Singularity Analysis," and "The Plan to Singularity." It doesn't matter much, though, whether these specific plans hold up under detailed expert scrutiny; they serve as an accessible model for the process we're discussing.

Here we see conventional open-source machine intelligence, starting with industrial AI, leading to a self-rewriting seed AI which runs right into singularity takeoff. You'd have a machine that combines the brains of a human (maybe literally, in coded format, although that is not part of Yudkowsky's scheme) with the speed and memory of a shockingly fast computer. It won't be like anything we've ever seen on earth. It will be able to optimize its abilities, compress its source code, turn its architecture from a swamp of mud huts into a gleaming, compact, ergonomic office (with, as it were, a spa and a bar in the penthouse, lest we think this is all grim earnest).[19] Here is quite a compelling portrait of what it might be like, "human high-level consciousness and AI rapid algorithmic performance combined synergetically," to be such a machine:

> Combining Deep Blue with Kasparov . . . yields a Kasparov who can wonder "How can I put a queen here?" and blink out for a fraction of a second while a million moves are automatically examined. At a higher level of integration, Kasparov's conscious perceptions of each consciously examined chess position may incorporate data culled from a million possibilities, and Kasparov's dozen examined positions may not be consciously simulated moves, but "skips" to the dozen most plausible futures five moves ahead.[20]

Such a machine, we see, is not really human-equivalent after all. If it isn't already transhuman or superhuman, it will be as soon as it has hacked through its own code and revised it (bit by bit, module by module, making mistakes and rebooting and trying again until the whole package comes out right). If that account has any validity, we also see why the decades-long pauses in the timetables cited earlier are dubious, if not preposterous. Given a human-level AI by 2039, it is not going to wait around biding its time until 2099 before creating a discontinuity in cognitive and technological history. That'll happen quite swiftly, since a self-optimizing machine (or upload, perhaps) will start to function so much faster than its human colleagues that they'd simply be left behind, along with Moore's plodding law. A key distinguishing feature, if Yudkowsky's analysis is sound, is that we never will see HAL, the autonomous AI in the movie *2001*. All we will see during the transitional period is impersonal AI specialized to develop software. "Nor do we have, say, uploads developed in 2040 with the first social effects starting to become visible in 2060. The only visible sign of the Singularity would be the dumber-than-human AI industry. When the end arrived, it'd occur within a couple of days. Frankly, I think this plateau-and-breakthrough model makes a lot more sense than the slow-but-steady Singularities."[21]

Waiting for the end, creating a beginning

Since I don't know the true shape of the future any more than you do, I certainly don't know whether an AI or nanominted Singularity will be brought about (assuming it does actually occur) by careful, effortful design in an institute with a Spike engraved on its door, by a congeries of industrial and scientific research vectors, or by military ambitions pouring zillions of dollars into a new arena that promises endless power through mayhem, or mayhem threateningly withheld.

It does seem excessively unlikely that we will skid to a stop anytime soon, or even that a conventional utopia minus any runaway singularity sequel (*Star Trek*'s complacent future, say) will roll off the mechanosynthesizing assembly line.[22]

What could stop the Spike? There are several obvious candidates.

Nuclear warfare, large or even comparatively local, could slow the arrival of a singularity, as Vinge himself recognized (his fiction uses a nuclear exchange exactly to delay the Singularity, forestalling too much future shock in his readers). Bacteriological or viral warfare might have a similar terrible impact. Neither would be able to exterminate all humans, or even the majority. Fears of nuclear winter in a spasm exchange between two superpowers (and Russia still owns many of the Soviet Union's nuclear missiles) have abated, and it seems unlikely that a limited nuclear exchange would do more than slow the pace of scientific research. It is remotely possible that some terrorist nation or faction might use early wet nano to develop infections tailored to defeat all our antibiotics and erode our immune systems, killing most of us and many animals as well. These are hideous hazards we need to confront and guard against in any case, whether or not we regard the idea of the Spike as plausible or desirable.

Similarly, a near-Earth asteroid or passing comet could slam into the planet, with truly devastating effect. An impact capable of wiping out the dinosaurs and many species on land and in the oceans would not leave much of our civilization standing. Again, the probability is very small that this would happen inside the time window before we reach singularity. Indeed, early in 2000 it was announced that the number of these dangerous space rocks is much smaller than earlier estimates: between 500 and 1000 are thought to be in dangerous orbits, of which 322 are now identified.

Then there's sheer overpopulation, pressing hard on our resources and species diversity in a global climate made troublesome and unstable by greenhouse effect, ozone damage, acid rain, and other perturbations. These hazards, too, clearly need to be addressed urgently, on grounds of simple humanity as well as security and stability. But will they be dealt with? Some steps forward have been taken, others have stalled. Powerful special interests seek short-term profit or exercise of power despite the risks of delaying action. Could these pressures cause the collapse of the very mercantile economy needed to build the machines and techniques driving the Spike? Perhaps. But don't forget, those technical advances are not being made purely for their intellectual beauty or challenge, as cosmological discoveries are. The primary components of the Singularity's upward curve—increasing computer power, superior software able to improve its own next generation of code, genomic and other bioscience consolidation, molecular manufacture at ever-smaller scales—will attract investment and use because they are powerfully effective, competitive, and profitable. As these technologies arrive or mature, they will *solve* the problems that now seem intractable. Many of the current causes of resource wastage, pollution, and harmful interference with the ecosphere will become unspeakable horrors of the past, as slavery is nowadays.

Are there boringly obvious technical obstacles to a Spike? We have seen that certain techniques will saturate and pass through inflexion points, tapering off their headlong thrust. If the past is any guide, new techniques will arrive (or be forced into reality by the lure of profit and sheer curiosity) in time to carry the curves upward at the same continuing acceleration. If not? Well, then, it will just take longer to reach the Spike, perhaps decades or even centuries rather than years, but it is hard to see why progress in the necessary technologies would simply *stop*.

Well, perhaps some of these options will become technically feasible but remain simply unattractive, and hence bypassed. Dr. Russell Blackford is a lawyer, writer, and critic who has written interestingly about social resistance to major innovation.[23] He notes that while human-crewed exploration of Mars has been a technical possibility for the past three decades, that challenge has not been taken up. Video-conferencing is available but few use it, and videophones simply never arrived as a substitute for the ordinary phone (unlike the instant adoption of mobiles). While a concerted program involving enough money and with widespread public support could bring conscious AI about by 2050, he argues, it won't happen. Conflicting social priorities will emerge, the task will be difficult and horrendously expensive.[24]

Are these objections valid? AI and nano need not be impossibly hard and costly, since they will flow from current work hammered forward by Moore's law improvements, including powerful, vastly complex genetic algorithms suddenly easier to "evolve" on ever-faster machines. Missions to Mars, by contrast, offer no obvious social or consumer or even scientific returns beyond their simple breathtaking achievement. Good science can be done by remote vehicles. By contrast, minting and AI or IA will bring immediate and copious benefits to those developing them—and will get less and less expensive, just as desktop computers have.

What of social forces taking up arms against this future? We've seen the start of a new round of protests and civil disruptions aimed at genetically engineered foods, work in cloning and genomics, and lately at longevity and AI research. The Turning Point Project has spent over a million dollars on big advertisements attacking nanotech, AI, and genetic engineering. The only question is how effective the impact will be.

In 1999, for example, emeritus professor Alan Kerr, winner of the lucrative inaugural Australia Prize for his work in plant pathology, broadcast a heartfelt denunciation of the Greens' adamant opposition to new genetically engineered crops that allow insecticide levels to be cut by half. Some aspects of advancing technology, however, did alarm Dr. Kerr. He admitted to being "scared witless" by the thesis "that within a generation or two, science will have conquered death and that humans will become immortal. Have you ever thought of the consequences to society and the environment of such an achievement? If you're anything like me, there might be a few sleepless nights ahead of you. Why don't the greenies get stuck into this potentially horrifying area of science, instead of attacking genetic engineering with all its promise for agriculture and the environment?"[25] This sort of diversionary tactic, I fear, is shortsighted and liable to rebound. While it might arouse confused opposition to life extension and other beneficial ongoing research programs, it will lash back as well against any other technology ill-understood by the public—such as safer bioshields against insect predation of crops.

Cultural objections to AI, cryonics, and uploading will emerge, as venomous as yesterday's and today's attacks on contraception and abortion rights, or antiracism struggles. If opposition to the Spike, or any of its contributing factors, gets attached to one or more influential religions or cults, that might set back or divert the rushing current. Alternatively, careful, balanced study of the risks of general assemblers and autonomous artificial intelligence might result in just the kinds of moratoriums that Greens now urge upon genetically engineered crops and herds.

Given the time lag before a singularity occurs—at least a decade, and far more probably at least two or even five—there's room for plenty of informed specialist and public debate of this sort. Just as the basic technologies of the Spike will depend on design-ahead projects, so too we'll need a kind of think-ahead program to prepare us for changes that otherwise might, indeed, scare us witless. And of course, the practical, day-to-day impact of new technologies always conditions the sorts of social values that emerge in response to their arrival. Recall the subtle interplay between availability of the oral contraceptive pill and swiftly changing sexual mores, the easy acceptance of *in vitro* conception.

Despite possible impediments to the arrival of the Spike, then, I suggest that while it might be delayed, almost certainly it's not going to be halted. If anything, the surging advances I see every day coming from labs around the world convince me that we already are racing up the lower slopes of its curve into the incomprehensible.

In short, it makes little sense to try to pin down the future with any exactness. Too many strange changes are occurring already, with more lurking just out of sight, ready to leap from the equations and surprise us. True AI, when it occurs, might dash within days or months to SI (superintelligence), and from there into a realm of Powers whose motives and plans we can't even start to second-guess. Nano minting could go feral or worse, used by crackpots or statesmen to squelch their foes and rapidly smear us all into paste. Or sublime AI Powers might use it to the same end, recycling our atoms into better living through femtotechnology.

The single thing I feel confident of is that one of these trajectories will start its visible run up the right-hand side of the graph within ten or twenty years, and by 2030 (or 2050 at latest) will have put into question everything we hold self-evident. We will live forever; or we will all perish most horribly; our minds will emigrate to cyberspace, starting the most ferocious overpopulation race ever seen on the planet; or our machines will Transcend and take us with them, or leave us in some peaceful backwater where the meek shall inherit the Earth. Or something else, something far weirder and . . . *unimaginable*. Don't blame me. That's what I promised you.

Afterword

Somehow I've managed to avoid a regular 9-to-5 job for the past quarter century, so in a way I've been in training for the Spike, and for the yawning jobless gulf stretching into it during the next quarter or half century. And I'm here to tell you that it isn't hard to find gratifying things to do to fill in the time, even when you're deprived of the merriment of the office, farm, or factory.

One of the things I do all day is read books (and, increasingly, scads of wonderful stuff posted free on the Internet). I'm told that once they've put the gray horror of school behind them many people never pick up another book in their lives. Others might manage one or two a year: a thriller, a romance, a guide to better golf. Such deprivation would be almost as terrible to me as putting out my one functioning eye (don't ask). In the last fifteen or twenty years, I've gobbled down many hundreds of splendid books about various sciences (and as many about literature and critical theory). Even though I've immediately forgotten most of what I read, something strange happens to a mind exposed to so much organized information. You start to see the world in a different way—in a *lot* of different ways, actually, many of them slightly at odds with one another. It's the perfect recipe for approaching the Spike, that informational tsunami on the horizon, that authentic embarrassment of riches.

So I express my heartfelt thanks to the authors of those many books, writers whose works I've plundered in stitching my own together: Paul Davies and Steven Weinberg and Murray Gell-Mann, David Deutsch and Lee Smolin and Julian Barbour, Randolph M. Nesse and George C. Williams, M. Mitchell Waldrop and Roger Lewin and Stuart Kauffman, Dennis Overbye and Ed Regis, John L. Casti and Douglas R. Hofstadter and Daniel Dennett, John Gribbin and Stephen Rose and Lynne Margulis (hmm, the first woman on this random list), Carl Sagan and Jerome

Kagan, Marshall T. Savage and Robert L. Forward and K. Eric Drexler, Arthur C. Clarke and Howard Rheingold and Sherry Turkle and Jeremy Campbell, Brian Greene and Howard Gardner and Stephen Jay Gould, and— Once started, you can't stop, but you have to. Above all, two very bright guys who built the spine of my book for me, Vernor Vinge and Hans Moravec. I commend them all, for the joy they'll bring you.

I would have stared less fiercely into the Singularity without the guidance (and permission to quote their Internet writings) of many intriguing people on the Extropian newsgroup, especially John K. Clark (Pelagius) and Lyle Burkhead (Augustine) who generously provided me with copious amounts of argumentative text posted before I had access to the mailings, Eliezer Yudkowsky, whose Web pages never stay still, and Eugene Leitl and Anders Sandberg, two extraordinary European transhumanists. Dr. Michael Nielsen, Robert J. Bradbury, Eliezer Yudkowsky, and Professor Robin Hanson read and commented helpfully on all or parts of the book. Gregory Burch kindly airmailed me an important document, and Hans Moravec sent me a copy of *Robot*. I'm grateful to all those who've permitted me to cite their remarks: Forrest Bishop, David Bofinger, Jonathan Burns, Dan Clemmensen, K. Eric Drexler, Thomas McCarthy, Ralph Merkle, Mitchell Porter, Gary Stix, Charles Stross, Robert Theobald, Wayne Throop, Paul Wakfer, and many others who have enlightened, enraged, and amused me during the writing of this complicated book.

Closer to home, I received helpful readings and advice from Paul Voermans, Russell Blackford, and Christina Lake, and useful pointers on nanotechnology from Geoff Leach at RMIT. I am especially grateful to Alan Stewart, who loaned me several out-of-print books, and to John Foyster and Chris Maxfield, who kindly sent me photocopies of crucial research material. As always in recent years, my access to the Internet and extensive library services have been provided by the University of Melbourne, where I'm a Senior Fellow of the Department of English and Cultural Studies. Finally, I thank my editor at Reed/New Holland,

Clare Coney, who commissioned the first edition of this book,
and David G. Hartwell, who bought the revised edition, enabling
me to gaze into the lurid and astonishing depths of the new
century and its Spike.

Melbourne, September 2000

Notes

1: The Headlong Rush of Time

1. *The New York Times Book Review,* 28 October 1984, pp. 1, 40–41; this citation from p. 41.

2. http:// www-rohan.sdsu.edu / faculty / vinge / misc /singularity.html or http://www.frc.ri.cmu.edu/~hpm/book98/com.ch1/vinge.singularity.html

3. It's like compounded interest. Say you're given $100 at birth, which is invested at 10 percent. Your earnings are plowed back into the capital so it also earns further interest. You'll double your original gift by the time you are seven years old, garnering a hundred extra bucks. If a more fortunate friend's godparents invest a million for her at the same rate, after the same seven years she's earned an extra million. When you're both in your sixties, you'll be worth around $70,000, but she'll own $70 billion—yet her rate of return is exactly the same as yours. Such is the power of exponential increase.

4. The likely costs and time of completion of such domestic durables is discussed briefly in the final, Summary chapter.

5. Robert J. Bradbury, private communication, 13 January 2000.

6. http://www.rand.org/publications/MR/MR615/mr615.html As long ago as 1995, this Rand report noted: "A key observation is that a number of countries are engaged in some level of effort relevant to the foundations of molecular nanotechnology. Although the United States has many groups performing work related to nanotechnology and molecular manufacturing, there are several strong competitors and potential collaborators." By July 2000, the US government was paying attention to nano. The National Nanotechnology Initiative released a 144–page report on prospects and funding (nearly half a billion dollars) for the next five years. Ironically, pioneering nano analysts such as Eric Drexler, Ralph Merkle, and Robert Freitas appear nowhere in this report. See http://www.nano.gov/nni2.htm

7. Daniel G. Clemmensen has offered a formal definition, while

stressing that we deviate here into analogy (and if your eyes skid over the equation, calm down and take another look). "It is a point on a function space where the function ceases to have meaning, usually because the function tends to infinity as it approaches the point. For example, $F(x) = 1/x$ has a singularity at $x = 0$." That is to say, when x is set equal to zero, 1 divided by x becomes *infinite*. Why? Well, suppose x equals something quite tiny, rather than nothing at all—say 0.000,000,000,000,000,001. It would take 1,000,000,000,000,000,000 of those teeny little x's to make up 1. It follows, loosely speaking, that when you increase the number of zeroes out to infinity, you are packing an infinite number of zeroes into 1.

8. Personal communication, 19 August 1996.

9. On the last day of 1999, Vinge told me in an e-mail: "The basic argument hasn't changed."

10. http://www.zyvex.com/PressReleases/MerkleJoinsZyvex.html.

11. Gregory S. Paul and Earl Cox, *Beyond Humanity: Cyberevolution and Future Minds,* 1997. My quotations from Paul and Cox are drawn from several key chapters posted by them on the World Wide Web. Their book received the *Choice* Award for Outstanding Academic Book for 1997.

12. A salient sample is at http://www.obs-us.com/english/books/rawlins/moths/unknown/9.html

13. Vinge's NASA Symposium paper on the Singularity cites this memoir thus: "One conversation centered on the ever accelerating progress of technology and changes in the mode of human life, which gives the appearance of approaching some essential singularity in the history of the race beyond which human affairs, as we know them, could not continue."

14. Barry Jones, *Sleepers, Wake!,* 1995 edition, p. x.

15. Barry Jones, in a review of the 1997 Australian edition of this book, linked from the Web at http://dargo.vicnet.net.au/ozlit/writers.cfm?id=74.0

16. G. Harry Stine, "Science Fiction Is Too Conservative," *Analog Science Fact & Fiction,* May 1961, pp. 84–99.

17. Stephen Jay Gould has explained what is *really* going on in sports record-breaking—the closing up of variance around the mean as ex-

cellence becomes more general and human limits are approached—in his powerfully argued *Life's Grandeur*.

18. G. Harry Stine, "Science Fiction Is Still Too Conservative," *Analog*, January 1985, pp. 89–96. I asked Stine for his current assessment. Without commenting on his wildly erroneous earlier projections, he replied gamely: "Science fiction is STILL too conservative!" (Personal communication, 1996).

19. BT 1997 projections can be found at http://btlabs1.labs.bt.com/library/on-line/calendar/index.htm. Oddly, this now slightly outdated list has not been updated.

20. Brian Alexander, "Don't Die, Stay Pretty," *Wired*, January 2000, p. 181.

21. It was beaten to the finish line by its commercial competitors, notably Craig Venter's Celera, which announced completion in mid-2000. See http://www.celera.com/corporate/about/press__releases/celera 062600__l.html

22. Richard Neville, "Fasten Your Seat Belts," Age *Good Weekend* Millennium issue, 1 January 2000 (Melbourne, Australia), p. 13.

23. See "Telomeres, Telomerase, and Cancer," by Carol W. Greider and Elizabeth H. Blackburn (who first cracked the telomere code in 1978), in *Scientific American*, February 1996, pp. 80–85. Blackburn was also the scientist who showed that telomere length fluctuates through a lifetime.

24. This topic is usefully discussed in "The Death of Old Age" by David Concar, *New Scientist*, 22 June 1996, pp. 24–29.

25. Response to the present author's contribution to the 1998 online Symposium on the Singularity, which can be found at http://www.extropy.org//eo/articles/vi.html.

26. http://www.boston.com/dailynews/003/world/

27. Robert C. W. Ettinger, *The Prospect of Immortality* (New York: Doubleday, 1964; Sidgwick and Jackson, 1965.) A privately published version had appeared in 1962.

28. "The Technical Feasibility of Cryonics," *Medical Hypotheses*, vol 39, 1992, pp. 6–16. A longer version of this paper is available on the Net at: http://merkle.com/merkleDir/techFeas.html

29. Freezing heads is a suggestion I published myself in 1971, some

years before it was put into practice by Alcor Life Extension Foundation in Riverside, California, a leading cryonics firm now in Scottsdale, Arizona. I've learned to my chagrin that I was anticipated in the late 1960s—by, sad to say, an early cryonics enthusiast who perished in a sailing accident, far from the necessary preserving equipment. There is no *absolutely* safe way to cheat death—although that is a stupid phrase, taken literally: there is, we must remind ourselves, no such thing as Death with designs upon us.

30. I am grateful to Dr. Mike Perry of Alcor for this estimate (personal communication 16 January 2000). Alcor stores 37 of these people, Ettinger's Cryonics Institute has 31, CryoSpan 14, and Trans Time two. Perhaps several others are preserved in private facilities.

2: Most Everything for Free

1. Even prior to Gibson's cyberspace, VR was predicted in some detail in Vernor Vinge's much-hailed 1981 short novel "True Names."

2. http://www.americanheart.org/Whats_News/_News_Releases/12-30-99_3-comment.html

3. Prospects for these technologies are described and discussed in Michael Dertouzos's *What Will Be: How the New World of Information Will Change Our Lives* (San Francisco: HarperEdge, 1997), pp. 68–77, 165–74. Dertouzos headed MIT's Laboratory for Computer Science for more than a quarter of a century.

4. See, for example, http://www.informatik.umu.se/~jwworth/medpage.html and http://www.umm.edu/news/releases/stealth.html

5. http://www.nlm.nih.gov/research/visible/

6. Cited by Howard Rheingold, *Virtual Reality* (1991), p. 168.

7. http://www.sandia.gov/media/NewsRel/NR1999/biosim.htm

8. http://www.gtri.gatech.edu/res-news/TGAME.html

9. http://www.best.com/~ddfr/Academic/Strong_Privacy/Strong_Privacy.html

10. Arthur C. Clarke, "The Lion of Commare," in *The Lion of Commare & Against the Fall of Night* (London: Corgi, 1975), p. 48.

11. Gregory Stock, *Metaman: Humans, Machines, and the Birth of a Global Super-organism* (1993), pp. 10, 14, 229; the citation below is from

p. 81. His enthusiasm for this totalizing, somewhat totalitarian prospect, reminiscent of Jesuit Teilhard de Chardin's "Omega Point," is perhaps surprising, given Stock's apparent anarcho-capitalist politics: he is an extropian, and has spoken at various of their conferences.

12. http://www.sunday-times.co.uk/news/pages/sti/2000/01/09/stinwenws 02004.html

13. Freeman J. Dyson, *The Sun, the Genome, & the Internet* (1999), p. 69.

14. http://hotwired.lycos.com/collections/space_exploration/6.02free-man_dyson_pr.html

15. This celebrated paper is available at http://www.zyvex.com/nanotech/feynman.html

16. Cited in "Foresight Update 20," available at the Foresight Institute Web site.

17. Cited from the 1998 virtual symposium *Comments on Vinge's Singularity*, organized by Dr. Robin Hanson and sited at Extropy's On-line URL.

18. See Ed Regis's splendid *Great Mambo Chicken* (p. 279) for the gory details.

19. Chris Peterson comments: "The book is copyrighted 1978. Eric had already thought of nanotechnology by then, but probably the reason that *Nanosystems* didn't mention Santa Claus Machines is that, as you point out earlier, 'That was distinctly not nanofacture.' It's not clear in what sense the concept is a predecessor" (personal communication, 17 March 1997). I do not dispute Drexler's priority with molecular nano-technology, of course, but it is interesting that the notion of self-assembling and even self-replicating von Neumann machines was current among NASA planners even as he hatched his own brilliant variant.

20. This discussion was posted on the transhumantech newsgroup in January 1998, and is cited with the authors' permission. Similar and more elaborate discussions occur on newsgroups such as sci.nanotech.

21. Personal communication, 15 January 2000.

22. In the extropian on-line symposium on Vinge's Singularity.

23. *Unbounding the Future,* pp. 153–59.

3: Little Things

1. Available at http://www.businessweek.com/1999/99_35/b3644007.htm

2. *Engines of Creation.*

3. John Barnes, "How to Build a Future," *Analog,* March 1990, vol 110, number 4.

4. Enunciated in *Profiles of the Future,* p. 36.

5. Documented, Bradbury notes, in Robert Freitas's *Nanomedicine,* vol I (1999), p. 295.

6. Personal communication, 13 January 2000.

7. See details of Tierra at http://www.hip.atr.co.jp/~ray/tierra/tierra.html

8. Cited in Kelly, *Out of Control.* See http://www.well.com/user/KK/Out of Control/

9. "Calculating with DNA," *Scientific American,* September 1995, p. 20.

10. Cited from the on-line extropian symposium on Vinge's Singularity. Bostrum is cofounder of The World Transhumanist Association (http://www.transhumanism.com), on the editorial board of *The Journal of Transhumanism* (http:/www.transhumanist.com) and cowriter of *The Transhumanist FAQ* (linked at his Web site).

4: Minting the Spike

1. http://www.well.com/conf/mirrorshades/

2. "Storrs is a family name that has been the middle name of the first born son for six generations. The University of Connecticut (at Storrs, Conn.) has a building named Storrs Hall. But they don't have a J." (Personal communication.)

3. Browse these lists and certain names start to jump out at you— Anders Sandberg (28, Swedish neuropsychology graduate student and intellectual omnivore, custodian of an important Transhumanism Net site); Eugene, or 'gene, Leitl (34, Russian-born German cryobiology researcher with degrees in chemistry, currently in California); John K. Clark (51, electrical engineer and "terminal bookworm"); Forrest

Bishop (nanotechnology designer; aerospace engineer and inventor of an airborne Remotely Piloted Vehicle); Mitch Porter (30, unemployed Australian B.Sc. generalist and perpetual student, described by the brightest person I know as the brightest person he knows); Daniel G. Clemmensen (50, Virginia specialist in distributed processing, computer communications, and multiprocessor architectures); Lyall Burkhead (54, West Coast engineer and "genie"-skeptic with a riddling Socratic style); Amara Graps (38, astronomer, science writer, and computational physicist studying interplanetary dust dynamics at the Max Planck Institute for Nuclear Physics in Heidelberg, Germany); Eric Watt Forste (34, World Wide Web developer and executive of the Extropian Institute); Natasha Vita More (in a previous life Nancie Clark but now an L.A. Transhumanist artist, author, and Automorpher); Max More, Ph.D. (36, philosopher, Dr. Extropian himself); Greg Burch (43, Texan attorney, wide-span intellectual and Extropian vice-president, who shares his home with lemurs); Robin Hanson, Ph.D. (41, economics professor at George Mason University and, like many of the others, impressive polymath, previous lives including computer research at the Lockheed Artificial Intelligence Center from 1984–89 and Bayesian statistics research, 1989–93); Corwyn J. Alambar (28, network administrator in Seattle, aiming someday to recast his body into "a large, predatory felinoid"); Mark Crosby (45, a Washington, D.C., systems specialist with the U.S. Consumer Price Index); at least one eager-beaver schoolkid (here nameless, for the protection of the innocent) who wags his transhuman tail like a friendly puppy dog, and you can't really be cross if sometimes he pees on your leg in his excitement; Twirlip of Greymist, a.k.a. The Low Golden Willow (a.k.a. Damien R. Sullivan of CalTech); QueeneMUSE (a.k.a. Nadia Reed, a.k.a. Raven St. Crow, 43, painter and multimedia artist living in Hawaii and Southern California); Robert J. Bradbury (44, another polymath with skills in computing, biochemistry, and longevity research, combined in the Aeiveos Corporation, of which he is president); Gina "Nanogirl" Miller (creator of a news service, Nanotechnology Industries); and one genuine former child prodigy, Eliezer Yudkowsky (born 11 September 1979; and when I first met him online at age 17, as articulate, imaginative, well read, and pushy as a teenaged Isaac Asimov). I took Yudkowsky's posted biographical details on faith, conscious as always that, as the famous *New Yorker* cartoon told us, "In cyberspace, nobody knows you're a dog." For all I knew, Eliezer might really have been a well-read, intellectually nimble 71-year-old Brooklyn grandmother. Since then I've met him and I need no further persuasion.

4. The Extropian Principles have gone through several revisions over the years. Some of the earlier formulations with a misleadingly sophomoric twang to them—Boundless Expansion, Dynamic Optimism—have now been rephrased. The seven principles are by no means absurd, given the extropian analysis of technological impacts on human and posthuman lives: perpetual progress, self-transformation, practical optimism, intelligent technology, open society, self-direction, and rational thinking. Dr. More's careful elaboration of these principles can be found at the Extropian Web site: www.extropy.org.

5. http://www.geniebusters.org/index.htm

6. "Augustine" in real life is Lyle Burkhead, who presents himself as a "creedal minimalist," sexual bondage enthusiast, and teasingly Socratic teacher. While his reflections on nanotechnology are often bracing and informed, some of his other opinions are dubious in the extreme. For example, he has questioned the reality of the Nazi gas chambers "at Auschwitz or Birkenau," perhaps as a device to shake up people who accept such reports without seeking the evidence for themselves. In fact, his own views appear to match the usual tawdry revisionist nonsense, including such disgraceful statements as "Those of you who are fascinated by 'memetic engineering' would do well to consider the Holocaust meme as the canonical example. If you can design a meme like that, you can take over the world . . . for a while." (28 December 1996). Needless to say, I cite his views on nanotechnology despite, and not because of, this sort of libel. His full case against "Santa Claus" nanotech is on the Web as "Nanotechnology without Genies—A Critique."

7. "Pelagius" is my nickname for John K. Clark, the Florida electrical engineer cited earlier.

8. Markus Krummenacker, "Steps Toward Molecular Manufacturing," published in 1994 in *Chem. Design Autom. News* and updated in 1996, available on-line at http://www.n-a-n-o.com/nano/cda-news/cda-news.html

9. "Open source" is the term introduced by Chris Peterson, Eric Drexler's associate and wife, to replace "free software," the somewhat

misleading term favored by Richard M. Stallman. The originator of GNU (an early freeware version of the proprietary Unix operating system), Stallman has remarked: "The term 'free software' is sometimes misunderstood—it has nothing to do with price. It is about freedom" to run and modify a given program, and "to redistribute copies, either gratis or for a fee." See: http://www.gnu.org/gnu/thegnuproject.html

In other words, whatever it's called, this form of software development does not preclude commercial distributions or paid support and documentation services. Neither will AI or nanotechnology development, even when accelerated by the contributions of enthusiasts.

10. "The Cathedral and the Bazaar": http://www.tuxedo.org/~esr/writings/cathedral-bazaar/cathedral-bazaar.html

"Homesteading the Noosphere": http://www.tuxedo.org/~esr/writings/homesteading/

"The Magic Cauldron: Indistinguishable From Magic": http://www.tuxedo.org/~esr/writings/magic-cauldron/

11. http://www.mp3.com/

12. Jones, 1995, pp. 104–5. Strictly speaking, the claim that "in economic history there is no remote equivalent" is flatly wrong. Economics professor Robin Hanson notes: "Standard economic growth *is* exactly such an exponential rise and fall." Presumably Jones means that only with miniaturization do we see the effect quite so starkly and swiftly.

13. Ibid, pp. 105–6.

14. http://www.nas.nasa.gov/Groups/SciTech/nano/index.html

15. http://www.ioppublishing.com/Journals/Catalogue/NA

16. http://ccf.arc.nasa.gov/dx/basket/storiesetc/Nanopix.html

17. http://www.nas.nasa.gov/~globus/papers/NanoSpace1999/paper.html

18. Their paper "Atomistic Design and Simulations of Nanoscale Machines and Assembly" is at http://www.wag.caltech.edu/gallery/nano_comp.html and some cute stills and movies are at http://www.wag.caltech.edu/gallery/gallery_nanotec.html

19. On the other hand, *Physics Today* noted in December 1999: "Terms of employment for research physicists are changing rapidly. Over the last ten or fifteen years, basic research groups at, for example, AT&T, IBM, and the petroleum companies have mostly been dismantled to be replaced by more applied and directed efforts. Even product-

related research groups have often shrunk. Scientists at all levels have been replaced, retired, terminated, moved, or fired—sometimes gracefully, sometimes not." See: http://www.aip.org/pt/kadanoff.htm

However, IBM and Microsoft are working together to make software applications using quantum states, about as rarefied as you can get. There is a payoff for the companies, though: they hope to create software that can only be used once. See: http://www.eetimes.com/story/OEG19991124S0047

And IBM keeps announcing startling innovations. They have a thin, cheap, flexible transistor that can be sprayed onto plastic and could be the basis for a large portable computer display you can roll up and tuck in your pocket. See: http://cnn.com/TECH/computing/9910/29/science.transistor.reut/index.html

20. These remarks are cited and paraphrased with permission from posts to the Transhuman e-mail list, 16 January 2000.

21. http://www.zyvex.com/index.html

22. http://www.zyvex.com/Zybot/Zybot.html

23. http://www.ntechcorp.com/

24. http://www.techreview.com/articles/nov98/whitesides.htm

25. This is the point where it would be easy to numb you with details and figures, the kind Carl Feynman helped JoSH work through, but I'll leave you to find his paper and read it on your own time. It's available on the Net, of course, like so many of the most interesting documents exploring the nano future.

26. Both these citations are drawn from the on-line Symposium on Vinge's Singularity.

5: Climbing the Slope

1. Vinge cites this as Ulam, S., "Tribute to John von Neumann," *Bulletin of the American Mathematical Society,* vol 64, number 3, part 2, May 1958, pp. 1–49.

2. Cited by Vinge as Good, I. J., "Speculations Concerning the First Ultraintelligent Machine," in *Advances in Computers,* vol 6, Franz L. Alt and Morris Rubinoff, eds., pp. 31–88, 1965, Academic Press.

3. Asimov, *A Choice of Catastrophes.*

4. Clemmensen adds, for the computer literate: "He's using LINUX, so he has the source."

5. A detailed examination of how this technology might develop in the next few decades can be read at Dr. Joe Strout's University of North Carolina, Chapel Hill, Web site: http://metalab.unc.edu/jstrout/uploading/MUHomePage.html

Here's an example of his analysis of upload demographics: "If people live a two-stage life cycle (i.e., live and have children in biological form, then upload as the body starts to fail), then uploading will not slow the growth of Earth's population. In fact, it will probably accelerate it. However, artificial bodies may be built which are adapted to other environments. They could live comfortably on the surface of the moon or in unpressurized orbiting habitats without protection. It will probably be uploaded pioneers who colonize the solar system, and eventually other stars, returning to Earth for the occasional visit.

"As population continues to grow, more and more people may choose to live in artificial realities, which can be much roomier on the inside than on the outside. One can imagine a great orbiting computer, a cubic kilometer of circuitry, housing billions of uploaded people in relative comfort. Or, perhaps, people will live instead in a great network of smaller computers, transferring themselves from one to another just as we send email around the Internet today."

6: Uncoupling the Flesh

1. This article can be found at: http://hanson.gmu.edu/uploads.html

2. See "Mind switch could help disabled regain control," by Ian Anderson, *New Scientist,* 4 May 1996, p. 6.

3. http://news.bbc.co.uk/hi/english/sci/tech/newsid_606000/606938.stm

4. http://www.wired.com/news/news/technology/story/22116.html

5. Vinge, *Threats . . .*

6. A fine sketch of this debate, and its principal disputants, is in the *New Scientist* report—"Zombies, Dolphins and Blindsight," 4 May 1996, pp. 20–7—by Alun Anderson, Bob Holmes, and Liz Else on the second biennial Tucson conference "Towards a Science of Consciousness." It is archived at http://www.newscientist.com

7. David Chalmers, *The Conscious Mind* (New York: Oxford University Press, 1996), pp. 314–5, 332.

8. A philosopher, Dr. Aubrey V. Townsend, discusses this matter of uploads and identity, and related matters, in an interesting paper, "Survival in Cyberspace," in the British journal *Foundation* (Summer 2000). He draws upon such philosophically complex theorists as Bernard Williams and Derek Parfit. See Williams, "The Self and the Future," in *Problems of the Self* (Cambridge: Cambridge University Press, 1973) and Parfit in *Reasons and Persons* (Oxford: Clarendon Press, 1984). Parfit is widely acknowledged as the great current theorist of this broad topic.

9. This blood-chilling possibility is the basis of James Patrick Kelly's "Think Like a Dinosaur," a bitterly sardonic revisionist take on a celebrated 1954 story by Tom Godwin, "The Cold Equations." There, a girl stowaway on a spacecraft bearing emergency medical supplies is ejected into space, because her unanticipated mass would cause the craft to crash. Kelly's dinosaurs are a haughty race of aliens who provide superluminal wormhole transport to humans, but insist that the "redundant duplicate"—that is, the person who is scanned before that datastream is sent across light-years and reassembled into the transported version—must be killed to protect universal harmony and balance. According to many fans of uploading, this would not be murder and perhaps not even a crime (as amputation without medical need might be). Kelly makes it shockingly clear that it is indeed murder, and most foul.

10. They do, however, in Norman Spinrad's interesting short novel *Deus X* (New York: Bantam Spectra, 1993), which hangs on this very point. And in Greg Egan's brilliant upload novel *Permutation City* (London: Millennium, 1994), uploads have a distressing way of killing themselves the moment they understand that they are the copies in a VR world and not the originals.

11. Calvin's commentary in the same issue, "Cautions on the Superhuman Transition," is also included in the closing chapter of his 1996 book *How the Brain Works*.

12. The details of Hanson's socioeconomic analysis can't be pursued here, but are eminently worth consulting. His URL is given at the end of this book.

13. It's rather like Kurt Vonnegut's gag in *Slapstick, or Lonesome No*

More! in which the Chinese communists shrink themselves to micro-scopic dimensions to avert the resource depletions of overpopulation. Blown in the wind, the "teeny-weeny Chinese" turn into the Green Death, caused by inhaling or ingesting them. Hanson's version is less macabrely funny than that, but has its own appeal.

14. The following Bartlett experiment is described in Jeremy Campbell, *The Improbable Machine: What the Upheavals in Artificial Intelligence Research Reveal About How the Mind Really Works* (New York: Simon & Schuster, 1989), pp. 91–2.

15. Gregory Benford, *Sailing Bright Eternity* ([1995] London: Vista, 1996), p. 405. I should make it clear that Dr. Benford puts these notions forward in works of fiction. But he also takes them quite seriously as speculative possibilities for our future (interview, September 1999).

7: The Interim Meaning of Life

1. W. French Anderson, "Gene Therapy," *Scientific American*, September 1995, p. 96.

2. Reported in the London *Sunday Times*: http://www.sunday-times.co.uk/news/pages/sti/1999/12/26/stinwenws02012.html

3. http://www.seattletimes.com/news/health-science/htm198/fetu_19990323.html

4. In their book *Beyond Humanity* (see bibliography).

5. *The Age*, 26 September 1996, p. A3.

6. Karl Zinsmeister, "Last Hurrah for Class Warriors," *Wall Street Journal*, 25 January 1996.

7. *Wall Street Journal*, 17 April 1996.

8. Personal communication, March 1997.

9. See "Measuring Genuine Progress" by Chris Nelder, on the worldwide green activist site at http://www.betterworld.com/BWZ/9610/learn.htm

A Canadian approach is presented by the Centre for the Study of Living Standards, at http://www.iosphere.net/~csls/news-01.html

10. Michael J. Mazarr, *Global Trends 2005*, p. 39.

11. Here are two recent analyses of the *benefits* of the proposal, vis-à-vis other distributional methods:

"An Efficiency Argument for the Guaranteed Income," Karl Widerquist and Michael A. Lewis (1998), which argues that the Guaranteed Income is the most efficient and comprehensive policy to address poverty: http://netec.ier.hit.u.ac.jp/WoPEc/data/Papers/wpawuwpma9802005.html

"Reciprocity and the Guaranteed Income," by Karl Widerquist (1998), which in part "addresses the criticism that the guaranteed income exploits middle-class workers by demonstrating that a basic income will have a positive effect on wages, which will at least partially counteract the effect of the taxes needed to pay for it." http://netec.ier.hit-u.ac.jp/WoPEc/data/Papers/wpawuwpma9808009.html

12. Robert Theobald's work is listed at http://www.transform.org/transform/tlc/rtpage.html

13. The 26 March 1995 UK *Observer* cover story "The Tomorrow People" reported Max More's opinion on privatizing the air. "Max's answer to worries about the environment advanced by eco-activists (great enemies of the Extropian struggle who, he says, often deliberately exaggerate visions of doom for their own purposes): 'We do need to be concerned about environmental issues . . . It's not our children and grandchildren we're worried about. It's ourselves.' Things might be helped, he adds, if resources were privately owned. 'The oceans and the air—anybody can pollute them without having to pay for it. Turn those commons into privately-owned resources and there'd be more rational economic use of them.' "

14. B. A. Santamaria, "Welfare will never compensate for work," *The Weekend Australian,* 21–2 December 1996, p. 22.

15. From a "special issue on the future" of *Spin* magazine guest-edited by Jaron Lanier, November 1995. Quoted with his permission from Lanier's Web homepage, http://www.well.com/Community/Jaron.Lanier/index.html

16. Thomas McCarthy, "MNT and the World System," at http://www.mccarthy.cx/WorldSystem/index.htm

17. http://www.nas.nasa.gov/Services/Education/SpaceSettlement/

18. On the World Wide Web, in revised form, at http://sysop mind.com/singularity.html

19. The term derives from the surgically modified supermouse and the once-moronic human Charlie, in Daniel Keyes's heartbreaking novel

Flowers for Algernon. Mouse and man perish from side effects, a narrative inevitability in such a tragedy and not, as Yudkowsky suggests, a consequence of a law of nature. His discussion is on the World Wide Web at http://sysopmind.com/algernon.html

20. "The Ethics of Cognitive Engineering" was actually written in response to an earlier edition of this book. It may be found on the World Wide Web at http://sysopmind.com/algernon_ethics.html

21. http://singinst.org/CaTAI.html

22. Quoted from *The Singularitarian Principles*, vol 2.

8: Earth Is but a Star

1. Cited from http://www.skeptic.com/990107.html
Tipler argues his case against accelerated expansion at http://www.journals.uchicago.edu/ApJ/journal/issues/ApJ/v511n2/38990/38990.html

2. John D. Barrow and Frank J. Tipler, *The Anthropic Cosmological Principle* (Oxford: Clarendon Press, 1986), pp. 676–7.

3. Mary Midgley, *Science as Salvation* (London: Routledge, 1992), pp. 195–211.

4. Frank J. Tipler, *The Physics of Immortality* (London: Macmillan, 1995), p. 1.

5. As a side effect of his bizarre theory, Tipler offered a prediction of the mass of the Higgs particle, shown in 2000 to be wrong.

6. Foes of Tipler's reductionist atheology will need to do more than wag fingers at his presumption. While the body of his enthralling book avoids explicit equations, there's a 123-page Appendix for Scientists where much of his advanced analysis is set out in mind-crushing detail. As well as copious reading in philosophy and theology, Tipler draws on material that "would require Ph.D.s in at least three disparate fields." His own doctorate is in global general relativity, and he can cope with the rest "only because I've spent the past 15 years teaching myself" particle physics and computer complexity theory. "I've done it," he adds cheerfully, "so you can do it." Don't bet on it.

7. http://xxx.lanl.gov/archive/e-print/astro-ph/9909143

8. Frederik Pohl's sequence of "Heechee" sf novels, launched with

Gateway, ventured upon this postulate. Much earlier, James Blish's *The Triumph of Time* did it with a sort of baby universe given shape by his immortal protagonist Amalfi's ontological codes. Stephen Baxter's recent sequence starting with *Timelike Infinity* also has a battle for the shape of the universal substrate, as does his *The Time Ships* (the authorized sequel to Wells's *The Time Machine*), and so does Paul McAuley's *Eternal Light*. It is becoming something of a cliché in contemporary sf, which loves to rewire the cosmos.

9. David T. R. Given informs me that in sf humorist Terry Pratchett's novel *Strata* "the *entire* universe is based on one form of intelligent life built on the remains of another form built on the remains of another, etc. If life had not formed during the initial 10^{-43} secs, the universe would be mere featureless matter."

10. I here reborrow the borrowed theological term for the ageless energy beings invented by James Blish in his 1962 novel *The Star Dwellers*.

11. The breadth and generosity of his speculations, coupled with their scientific rigor and the diligence of his research, make his 1999 book *Robot* a major advance in the hitherto-unknown discipline of. . . . Spike studies. This enthralling book was delayed for several years due to disputes with his publisher, before he shifted it to Oxford University Press. I am grateful to Dr. Moravec for permission to read it in advance of publication.

12. This ludicrous state of overproduction was long ago satirically dubbed the "Midas Plague" by Frederik Pohl.

13. I discuss this speculative large-scale cyclicity in cultural history in *Theory and Its Discontents* (Melbourne: Deakin University Press, 1997).

14. It's a neat coinage, although I don't believe it will ever catch on; it's slightly too clumsy on the tongue.

15. Posted to the extropian e-list 15 October 1996, and cited with Mr. Witham's permission.

16. See, for example, Robin Hanson's reply on the extropian e-list, 15 October 1996.

17. "The New Cosmogony," in his delightful collection of reviews of nonexistent books, *A Perfect Vacuum* (London, Mandarin, 1991)— originally in English in 1979, sublimely translated by Michael Kandel— pp. 197–229. I am grateful to Mitch Porter and John Redford for reminding me of this wonderful, funny piece.

18. http://www.aeiveos.com/~bradbury/MatrioshkaBrains/

19. Yes, calm down, I know what I'm doing. While I continue to insist that religion, regarded literally, is the wrong interpretive filter to place over the Singularity, still the iconographies of a millennium of richly embroidered sacred art do yield a suitable set of metaphors for the strictly unimaginable, as the Marxist social and literary theorist Fredric Jameson noted more than once. And no, the Spike really *won't be the Rapture*. On the other hand, it has a far higher likelihood of taking place.

20. Julian Cribb, "Eco-disasters part of the new Russian Gulag." *Weekend Australian,* 2–3 March 1996, p. 48.

21. "Team Canada 1999 Country Market Report": http://www.dfait-maeci.gc.ca/english/geo/europe/40215-e.htm

22. A case argued by the late Julian Simon in *The Ultimate Resource,* at http://www.inform.umd.edu/EdRes/Colleges/BMGT/.Faculty/JSimon/Ultimate_Resource/

23. In his response to the 1998 on-line extropian symposium: "The notion of 'wall' or 'prediction horizon' is part of what I wanted to convey with the term 'singularity' (i.e., by metaphor with the use of the term in general relativity, which is no doubt from the use in math and mathematical modelling, the idea of a place where a smooth simple model fails and some more complicated description is necessary). In particular, I don't claim that the technological singularity implies that anything becomes infinite."

Summary: Paths and Time-lines to the Spike

1. See, for a simplified discussion, Nobelist Steven Weinberg's summary article "A Unified Physics by 2050?," *Scientific American,* December 1999, pp. 36–43.

2. http://www.wired.com/wired/archive/8:04/joy.html

3. Again, see Julian Simon's readable and optimistic *The Ultimate Resource.*

4. http://www4.nationalacademies.org/news.nsf/isbn/0309068916?OpenDocument

5. Stanislaw Lem, "Golem XIV" in *Imaginary Magnitude* (London: Mandarin, 1991).

6. Personal communication, 14 January 2000.

7. Michio Kaku, *Visions* (1998), p. 28.

8. http://news.bbc.co.uk/hi/english/sci/tech/newsid_503000/503552.stm

9. Michael Dertouzos, *What Will Be: How the New World of Information Will Change our Lives* (San Francisco: HarperEdge, 1997), p. 134.

10. Personal communication, 8 December 1999.

11. Personal communication, 16 January 2000.

12. Vernor Vinge, "The Digital Gaia," *Wired*, January 2000, pp. 74–8.

13. http://www.nas.nasa.gov/Groups/SciTech/nano/index.html

14. In a discussion on the extropian e-mail list, in the thread "Understanding Nanotech" begun on 26 August 1999.

15. Robert Freitas, *Nanomedicine*, vol I, 1999, p. 175.

16. Some thoughts on the difficult of containing nanotech (with some comparisons to software piracy "warez"), and the likely evaporation of our current economy, can be found in: http://www.cabell.org/Quincy/Documents/Nanotechnology/hello_nanotechnology.html

17. Personal communication, 8 December 1999.

18. A recent paper by Eric Drexler, "Building Molecular Machine Systems" in *Trends in Biotechnology*, January 1999, vol 17 no 1, pp. 5–7, is at http://www.imm.org/Reports/Rep008.html

19. Sorry, that's me again; Yudkowsky didn't say it.

20. http://singinst.org/CaTAI.html

21. Personal communication, 11 January 2000.

22. To be fair, the *Star Trek* franchise has always made room for alien civilizations that have passed through a singularity and become as gods. It's just that television's notion of post-Spike entities stops short at mimicry of Greek and Roman mythology (Xena the Warrior Princess goes to the future), spiritualized transformations of humans into a sort of angel (familiar also from *Babylon-5*), down-market cyberpunk collectivity (the Borg), or sardonic whimsy (the entertaining character Q, from the Q dimension, where almost anything can happen and usually does). It's hard not to wonder why immortality is not assured by the transporter or the replicator, which can obviously encode a whole person as easily as a piping hot cup of Earl Grey tea, or why people age and die despite the future's superb medicine. The reasons, obviously,

have nothing to do with plausible extrapolation and everything to do with telling an entertaining tale, using a range of contemporary human actors, that appeals to the largest demographic and ruffles as few feathers as possible while still delivering some faint frisson of difference and future shock.

23. Relevant articles from the Australian public affairs journal *Quadrant*: "Singularity Shadow," no. 347 (June 1998), "Women and Physics, Philosophy and Cyberspace," no. 357 (June 1999), "Genetics, Ethics, and the State," No. 359 (September 1999), "Life Extension and its Enemies," no. 362 (December 1999).

24. Blackford's case does not deny that we might live in a world of nanominted abundance by 2050, nor that by then there may be people who potentially have very long life spans. What he does not expect before 2100 is "lights-on" or conscious, volitional superintelligence, whether AI or upload (involving software more than hardware obstacles), which is required to send the technology curve pretty much vertical. If a true Spike of that kind happens, it would be delayed until these immortals in a world of abundance start to play. On this analysis, the utility of trying to produce these outcomes before about 2100 (efforts sure to be thwarted by their technical difficulty until that time) would operate as a drag on other *truly* radical change.

25. http://abc.net.au/rn/science/ockham/stories/s54399.htm

The book that frightened Dr. Kerr was my *The Last Mortal Generation*. The present book, by contrast, will surely shock him rigid.

Suggested Reading

Benedikt, Michael (ed). *Cyberspace: First Steps*. Cambridge, Mass.: The MIT Press, 1992.

Broderick, Damien. *The Architecture of Babel: Discourses of Literature and Science*. Melbourne: Melbourne University Press, 1994.

———. *Theory and Its Discontents*. Melbourne: Deakin University Press, 1997.

———. *The Last Mortal Generation: How Science Will Alter Our Lives in the 21st Century*. Sydney: New Holland, 1999.

Coveney, Peter, and Roger Highfield. *Frontiers of Complexity: The Search for Order in a Chaotic World*. London: Faber & Faber, 1995.

Barrow, John D., and Frank J. Tipler. *The Anthropic Cosmological Principle*. Oxford: Clarendon Press, 1986.

Calvin, William H. *The Cerebral Code: Thinking a Thought in the Mosaics of the Mind*. Cambridge, Mass.: The MIT Press, 1996.

Clarke, Arthur C. *Profiles of the Future*, rev. ed. London: Pan Books, 1999.

Damasio, Antonio R. *Descartes' Error: Emotion, Reason, and the Human Brain*. London: Picador, 1994.

Davies, Paul. *Superforce: The Search for a Grand Unified Theory of Nature*, rev. ed. London: Penguin, 1995.

———. *The Cosmic Blueprint: Order and Complexity at the Edge of Chaos*. London: Penguin, 1995.

Dennett, Daniel C. *Consciousness Explained*. Allen Lane, London: 1992.

———. *Darwin's Dangerous Idea: Evolution and the Meanings of Life*. London: Allen Lane, 1995.

———. *Kinds of Minds: Towards an Understanding of Consciousness*. London: Weidenfeld & Nicolson, 1996.

Drexler, K. Eric. *Engines of Creation: The Coming Era of Nanotechnology*. New York: Doubleday Anchor, 1986.

———. *Nanosystems: Molecular Machinery, Manufacturing, and Computation*. New York: John Wiley, 1992.

———, and Chris Peterson with Gayle Pergamit. *Unbounding the Fu-

ture: The Nanotechnology Revolution. New York: William Morrow, 1991.

Dyson, Freeman J. *The Sun, the Genome, and the Internet: Tools of Scientific Revolution*. Oxford: Oxford University Press, 1999.

Edelman, Gerald. *Bright Air, Brilliant Fire: On the Matter of the Mind*. London: Allen Lane, 1992.

Esfandiary, F. M. *Up-Wingers*. New York: John Day, 1973.

Freitas, Robert A., Jr. *Nanomedicine, Volume I: Basic Capabilities*. Austin, Tex.: Landes Bioscience, 1999.

Gould, Stephen Jay. *Life's Grandeur: The Spread of Excellence from Plato to Darwin*. London: Jonathan Cape, 1996.

Gribbin, John. *Schrödinger's Kittens and the Search for Reality*. London: Weidenfeld & Nicolson, 1995.

Hanson, Robin. "Is a Singularity Just Around the Corner?" *Journal of Transhumanism*, http://www.transhumanist.com/volume2/singularity.htm

Hapgood, Fred. *Up the Infinite Corridor: MIT and the Technological Imagination*. Reading, Mass.: Addison-Wesley, 1993.

Johnson, George. *Fire in the Mind: Science, Faith, and the Search for Order*. London: Viking, 1996.

Jones, Barry. *Sleepers, Wake!: Technology & the Future of Work*. Melbourne: Oxford University Press, [1982] 1995.

Kaku, Michio. *Hyperspace: A Scientific Odyssey Through Parallel Universes, Times Warps, and the 10th Dimension*. New York and Oxford: Oxford University Press, 1994.

————. *Visions: How Science Will Revolutionize the 21st Century and Beyond*. Oxford: Oxford University Press, 1998.

Kauffman, Stuart A. *The Origins of Order: Self-Organization and Selection in Evolution*. New York and Oxford: Oxford University Press, 1993.

————. *At Home in the Universe: The Search for the Laws of Complexity*. London: Penguin, 1996.

Kurzweil, Ray. *The Age of Spiritual Machines: When Computers Exceed Human Intelligence*. Sydney: Allen and Unwin, 1999.

Lem, Stanislaw (translated, Marc E. Heine). *Imaginary Magnitude*. London: Mandarin, 1991.

Lewin, Roger. *Complexity: Life at the Edge of Chaos*. London: Dent, 1993.

Mazarr, Michael J. *Global Trends 2005*. London: Macmillan, 1999.

McCrone, John. *The Myth of Irrationality: The Science of the Mind from Plato to Star Trek*. London: Macmillan, 1993.

Milburn, Gerard. *Quantum Technology*. Sydney: Allen and Unwin, 1996.

Moravec, Hans. *Mind Children: The Future of Robot and Human Intelligence*. Cambridge, Mass.: Harvard University Press, 1988.

————. *Robot: Mere Machine to Transcendent Mind*. Oxford: Oxford University Press, 1999.

Nesse, Randolph M., and George G. Williams. *Evolution and Healing: The New Science of Darwinian Medicine*. London: Weidenfeld and Nicolson, 1995.

Paul, Gregory S., and Earl Cox. *Beyond Humanity: Cyberevolution and Future Minds*. Rockland, Mass.: Charles River Media, 1997.

Penrose, Roger. *The Emperor's New Mind: Concerning Computers, Minds, and the Laws of Physics*. Oxford: Oxford University Press, 1989.

————. *Shadows of the Mind: A Search for the Missing Science of Consciousness*. Oxford: Oxford University Press, 1994.

Pinker, Steven. *The Language Instinct: The New Science of Language and Mind*. London: Allen Lane, 1994.

Regis, Ed. *Great Mambo Chicken and the Transhuman Condition: Science Slightly over the Edge*. London: Viking, 1991.

————. *Nano! Remaking the World Atom by Atom*. London: Transworld, 1995.

————. "Meet the Extropians," *Wired*, 2:10.

Rheingold, Howard. *Virtual Reality: Exploring the Brave New Technologies of Artificial Experience and Interactive Worlds from Cyberspace to Teledildonics*. London: Secker and Warburg, 1991.

————. *The Virtual Community: Finding Connection in a Computerized World*. London: Secker and Warburg, 1994.

Savage, Marshall T. *The Millennial Project: Colonizing the Galaxy in Eight Easy Steps*. Boston: Little, Brown, 1994.

Stableford, Brian, and David Langford. *The Third Millennium*. London: Paladin, 1988.

Stephenson, Neal. *The Diamond Age*. New York: Bantam Books, 1995.

Stix, Gary. "Trends in Nanotechnology: Waiting for Breakthroughs," *Scientific American*, April 1996, pp. 94–99.

Stock, Gregory. *Metaman: Humans, Machines, and the Birth of a Global Superorganism.* New York: Bantam, 1993.

Symposium on the Vingean Singularity: http://www.extropy.org/eo/articles/vi.html

Theobald, Robert. *Free Men and Free Markets.* New York: Doubleday, 1963.

————— (ed). *The Guaranteed Income: Next Step in Economic Evolution?* New York: Doubleday, 1966.

Thorne, Kip S. *Black Holes and Time Warps: Einstein's Outrageous Legacy.* London: Picador, 1994.

Tipler, Frank J. *The Physics of Immortality: Modern Cosmology, God, and the Resurrection of the Dead.* London: Macmillan, 1995.

Vinge, Vernor. *True Names . . . and Other Dangers.* New York: Baen Books, 1987.

—————. *Threats . . . and Other Promises.* New York: Baen Books, 1988.

—————. *Marooned in Realtime.* London: Pan Books, 1987.

—————. *A Fire upon the Deep.* London: Millennium, 1992.

—————. Address to NASA Vision-21 Symposium, March 30–31, 1993, downloadable from http://www-rohan.sdsu.edu/faculty/vinge/misc/singularity.html or http://www.frc.ri.cmu.edu/~hpm/book98/com.ch1/vinge.singularity.html

—————. "The Digital Gaia," *Wired.* January 2000.

Vita-More, Natasha. *Create: Recreate: The Third Millennial Culture*, 2d ed. Los Angeles: MoreArt, 1999.

Waldrop, M. Mitchell. *Complexity: The Emerging Science at the Edge of Order and Chaos.* London: Viking, 1993.

Selected Internet Sites

Bradbury, Robert J.: http://www.aeiveos.com/~bradbury/

Burch, Greg: http://users.aol.com/gburch1

Burkhead, Lyle: http://www.geniebusters.org/index.htm

Calvin, William H.: http://faculty.washington.edu/wcalvin/

Clemmensen, Dan: http://bobo.shirenet.com/~dgc/singularity/singularity.htm

Extropy Institute: http://www.extropy.org

Foresight Institute: http://www.foresight.org

Freitas, Robert: http://www.foresight.org/Nanomedicine/index. html

Hanson, Robin: http://hanson.gmu.edu

Kurzweil, Ray: http://www.kurzweiltech.com

Merkle, Ralph: http://merkle.com/merkle

Miller, Gina "Nanogirl": http://www.nanoindustries.com

Moravec, Hans: http://www.frc.ri.cmu.edu/~hpm/book98/

NASA Nano Team: http://www.nas.nasa.gov/Groups/SciTec/nano/
index.html

National Nanotechnology Initiative: http://www.nano.gov/nni2.
htm

Sandberg, Anders: http:/www.nada.kth.se/~asa/

Transhumanist Times: http://www.transhumanisttimes.com

Transhuman sites: http://www.anzwers.net/free/tech/websites.html
http://dmoz.org/Society/Philosophy/Current__Movements/
Transhumanism/Singularity

Vita-More, Natasha: http://www.extropic-art.com

Yudkowsky, Eliezer S.: http://singinst.org/beyond.html

Many other links relevant to the Singularity, its technologies, and
their expected impact, can be found by browsing these sites.

Index